About IFPRI

The International Food Policy Research Institute (IFPRI), established in 1975, provides research-based policy solutions to sustainably reduce poverty and end hunger and malnutrition. IFPRI's strategic research aims to foster a climate-resilient and sustainable food supply; promote healthy diets and nutrition for all; build inclusive and efficient markets, trade systems, and food industries; transform agricultural and rural economies; and strengthen institutions and governance. Gender is integrated in all the Institute's work. Partnerships, communications, capacity strengthening, and data and knowledge management are essential components to translate IFPRI's research from action to impact. The Institute's regional and country programs play a critical role in responding to demand for food policy research and in delivering holistic support for country-led development. IFPRI collaborates with partners around the world.

IFPRI

GLOBAL FOOD POLICY REPORT

2019

A Peer-Reviewed Publication

RECOMMENDED CITATION: International Food Policy Research Institute. 2019. *2019 Global Food Policy Report.* Washington, DC: International Food Policy Research Institute. https://doi.org/10.2499/9780896293502

International Food Policy Research Institute
1201 Eye Street, NW, 12th floor
Washington, DC 20005 USA
Telephone: +1-202-862-5600
www.ifpri.org

ISBN: 978-0-89629-350-2
DOI: https://doi.org/10.2499/9780896293502
Cataloging-in-Publication Data is available from the Library of Congress.

Photo credits
COVER: Anton Ivanov/Shutterstock.
CHAPTER IMAGES: p. 7 Adam Dean/Panos; p. 10 (background) Tommy Trenchard/Panos, (left) Sven Torfinn/Panos, (top) Patrick Brown/Panos; p. 11 Jenny Matthews/Panos; p. 17 Sven Torfinn/Panos; p. 27 G.M.B. Akash/Panos; p. 37 Tim Dirven/Panos; p. 45 Abbie Trayler-Smith/Panos; p. 53 Sven Torfinn/Panos; p. 61 Tim Dirven/Panos; p. 69 Sven Torfinn/Panos; p. 79 Martin Roemers/Panos; p. 87 Shanti Hesse/Shutterstock; p. 119 Abbie Trayler-Smith/Panos; p. 158 Jen Watson/Shutterstock.

BOOK LAYOUT: Jason Chow
EDITORIAL MANAGER: Pamela Stedman-Edwards

Contents

Preface

The *2019 Global Food Policy Report* reviews major food policy developments and events from the past year. In this eighth annual report, leading researchers, policymakers, and practitioners examine what happened in food policy in 2018 and why, and look ahead to 2019.

In 2018, many regions of the world faced increasing rates of hunger—with global undernourishment continuing to rise for the third year in a row—and stagnation in tackling malnutrition. A United Nations report found that conflict and climate change were key factors holding back countries' progress in achieving the Sustainable Development Goals (SDGs), though there has been some progress in reducing poverty and under-five mortality and in increasing access to electricity for populations in poor countries. Trade protectionism—especially evident in the trade war between the United States and China—and isolationism gained traction in many countries this year, with mixed impacts for agriculture, but negotiation of key trade agreements in Asia and Africa signaled a pushback to protectionism. Food and nutrition security remained largely on the periphery in many of the year's high-level meetings, such as the World Economic Forum and the 73rd session of the UN General Assembly. However, the international development community rallied support for SDG2–Zero Hunger, highlighted by the IFPRI-FAO global event on Accelerating the End of Hunger and Malnutrition, convened at the end of the year.

This year's report spotlights the urgent need for rural revitalization to address persistent crises in the world's rural areas. In recent years, political attention has shifted away from rural areas, and in many developing countries, limited rural employment opportunities, poor access to basic services, and worsening environmental degradation continue to contribute to persistent poverty, hunger, and malnutrition. A systemwide transformation is needed to revitalize rural areas—not only to achieve the SDGs, but more broadly, to make rural areas vibrant and healthy places to live and work.

There is reason for optimism. New opportunities and technologies and the expanding body of knowledge and evidence explored in this report can pave the way for reducing poverty and malnutrition and revitalizing the world's rural areas. Focusing on the needs of rural areas offers one of the most practical ways to achieve the SDGs and address many persistent and growing challenges. IFPRI joins with the United Nations Development Programme in the report's signature chapter to introduce rurbanomics, an innovative approach to rural development that builds on rural-urban synergies. Rapid urbanization around the world and growing urban demand for food and other products create new opportunities for rural people. Other chapters expand on the opportunities and requirements for rural revitalization. Among the most serious challenges facing rural areas is the lack of adequate employment opportunities: policies and investments should look to provide good jobs in the agrifood production and nonfarm sectors. Growth in employment and broader improvements in rural lives will depend on better provision of basic services—from roads to clean water to education. Falling prices for solar energy, for example, offer great promise for developing rural economies. Incentives to restore and improve rural environments can ensure valuable ecosystem services for both rural and urban residents and ensure the health of the planet. Women will play a central role in creating rural change—not only in addressing hunger and malnutrition but also by contributing to rural revitalization more broadly if access to skills, resources, and political voice is more equitable. Giving voice to all rural people is essential to ensuring that institutions and governance support real and sustainable improvements in rural lives. Experience in Europe with rural revitalization programs can help to inform similar efforts in developing countries. With political will and investment, we can tap into this vast potential to reverse the trends that proved so challenging in 2018.

Regional insights look at rural revitalization efforts, review events and trends in 2018, and provide a look ahead to 2019. Supplemented by data tables illustrating trends in key food policy indicators at the country and regional levels, the report provides a comprehensive overview of food policy.

I hope this report spurs action by policymakers as well as academics and members of the business community, civil society, and the media, all of whom have a stake in food policies that benefit the world's poorest and most vulnerable people.

SHENGGEN FAN
Director General

Acknowledgments

The *2019 Global Food Policy Report* was prepared under the overall leadership of Shenggen Fan and a core team comprising Jamed Falik, Rajul Pandya-Lorch, Pamela Stedman-Edwards, Klaus von Grebmer, Sivan Yosef, and Laura Zseleczky.

Text and data contributions were made by Akhter Ahmed, Kamiljon Akramov, Channing Arndt, Suresh Babu, Ousmane Badiane, Nienke Beintema, Clemens Breisinger, Kevin Chen, Emily Cho, Julia Collins, Stephen Davies, David Dawe, Eugenio Díaz-Bonilla, Xinshen Diao, Paul Dorosh, Jarilkasin Ilyasov, P. K. Joshi, Mekamu Kedir Jemal, Adam Kennedy, Homi Kharas, Nadim Khouri, Katrina Kosec, Anjani Kumar, Sarah Lowder, Fatma Mahfouz, Tsitsi Makombe, Hazel Malapit, Ian Masias, Alan Matthews, Ruth Meinzen-Dick, Hafiz Mirza, Alejandro Nin-Pratt, Lorenz Noe, Allen Park, Namita Paul, Valeria Piñeiro, Nilam Prasai, Agnes Quisumbing, Abdul Wajid Rana, Danielle Resnick, Claudia Ringler, Achim Steiner, Timothy Sulser, Timothy Thomas, James Thurlow, Peter Timmer, Evgeny Tsvetnov, John Ulimwengu, Zimeiyi Wang, Keith Wiebe, Sivan Yosef, and Laura Zseleczky.

Production of the report was led by Pamela Stedman-Edwards and Jamed Falik, with support from Jason Chow and Michael Go. Editorial assistance was provided by Tracy Brown, Amy Gautam, and John Whitehead.

The report underwent a peer review by IFPRI's Publications Review Committee.

CHAPTER 1

Food Policy in 2018–2019
Growing Urgency to Address the SDGs

SHENGGEN FAN

Shenggen Fan is director general, International Food Policy Research Institute, Washington, DC, USA.

2018 WAS A SOMBER AND UNPREDICTABLE YEAR, NOT ONLY FOR FOOD AND NUTRITION security, but also for global political stability and international development. Many regions of the world faced rising rates of hunger and stagnation in tackling malnutrition. Countries saw dramatic shifts in multilateral and bilateral relations. Trade protectionism and antiglobalism emerged as central themes in many national political landscapes, and foreign direct investment in food and agriculture suffered. Rural people around the world continued to face a crisis as they struggled with food insecurity, persistent poverty, and degraded land and water. In a rare bright spot, there was renewed emphasis on using entrepreneurial innovation and technologies to address development issues.

If the world is to achieve the Sustainable Development Goals (SDGs) by 2030, a fundamental transformation of our food and agricultural system and of rural areas is urgently needed. As this *Global Food Policy Report* highlights, rural revitalization is the linchpin of such a transformation. Revitalizing rural areas can stimulate economic growth and begin to address the crises in these regions, and also tackle both new and persistent challenges that are holding back achievement of the SDGs. As this report makes clear, rural revitalization is timely, achievable, and, most important, critical to ending hunger and malnutrition in just over a decade.

LOOKING BACK AT 2018

2018 painted a sobering picture of hunger and malnutrition. In September, the Food and Agriculture Organization of the United Nations (FAO) reported that undernourishment rose for the third year in a row. In 2017, 821 million people faced chronic food deprivation, as compared with 804 million in 2016 and 784 million in 2015.[1] These numbers reflect stagnation in many regions. Progress toward ending hunger in the Asia-Pacific region, for example, has stalled. Nearly half a billion Asians remain hungry, and urbanization does not appear to afford protection from hunger: in Bangkok, for example, more than one-third of children aged 6 to 23 months do not receive an adequate diet.[2]

The nutrition landscape offered some glimpses of slow progress. Globally, the number of stunted children declined by 9 percent from 2012 to 2017, and the share of infants under six months of age being exclusively breastfed rose slightly from 36.9 percent to 40.7 percent over the same period.[3] However, most other indicators point to a difficult road to achieving the 2030 development agenda. The prevalence of anemia among women of reproductive age rose from 30.3 percent in 2012 to 32.8 percent in 2016; no region in the world exhibited a decline, posing lifelong and generations-long problems for health and development.[4] The global proportion of overweight children remained relatively stable over the past five years (increasing from 5.4 percent in 2012 to 5.6 percent in 2017), but adult obesity continued to accelerate and worsen, with more than one in eight adults, or 672 million people, now considered obese.[5]

Rural areas continued to operate in crisis mode. Globally, 80 percent of the extreme poor (living on less than $1.90 per day) and 75 percent of the moderately poor (living on $1.90 to $3.20 per day) live in rural areas.[6] The prevalence of rural stunting is 26.8 percent, as compared with 19.2 percent in urban areas.[7] Rural people continued to suffer from a lack of economic opportunities and even basic services, such as water, sanitation, and electricity; and those living in conflict-affected countries found themselves in a vicious cycle of instability and food insecurity, prompting migration and displacement. Indeed, many of the challenges encountered this year disproportionately impacted rural people—from lack of healthcare in rural China, to the burden on rural women posed by India's worst water crisis in history, to rural Africans' increasing vulnerability to climatic shocks.

In June, the United Nations (UN) released its status report on the SDGs, which found that conflict and climate change are holding back countries' progress in achieving the global goals. But the report also highlighted some long-term positive trends that give cause for hope: the proportion of households living on less than $1.90 per person per day declined from 26.9 percent to 9.2 percent between 2000 and 2017. Over the same period, mortality of children under age five was slashed by almost 50 percent and the number of people in poor countries with access to electricity more than doubled.[8]

STRONG ECONOMIC GROWTH

Global economic growth remained strong and was projected to reach 3.1 percent in 2018, as it did in 2017. Although only 45 percent of countries were expected to experience further acceleration of growth in 2018, down from 56 percent in 2017, overall economic growth looked positive.[9] In South Asia, positive consumer and investor outlooks boosted growth from 6.6 percent in 2017 to 6.9 percent in 2018. Africa south of the Sahara saw 3.1 percent growth in 2018, up from 2.6 percent in 2017, reflecting a rise in oil and metals production, higher commodity prices, improving agricultural conditions, and booming domestic demand.[10] Africa also enjoyed a continuing reduction in the incidence and intensity of poverty, which fell by 4 percentage points between 2003–2008 and 2008–2017, while regionwide agricultural value added grew at 4.3 percent annually.[11] The Middle East and North Africa did not share in this good news: with the exception of Egypt, per capita economic growth rates remained modest and even negative in many countries, underscoring the challenges the region faces in addressing high rates of unemployment.

Global production of wheat, milled rice, and oilseeds in 2018 matched 2017's record production of most cereals.[12] Production of coarse grains decreased slightly (from 1,414 million metric tons in 2016/17 to 1,357 million metric tons in 2017/18).[13] Growth in global demand for agricultural commodities is slowing, reflecting the deceleration of China's per capita income growth that drove demand increases over the past decade. Rising demand from India and Africa south of the Sahara may help keep global demand growing.[14]

GLOBAL ATTENTION TO FOOD AND NUTRITION SECURITY

Food and nutrition security were only peripheral issues in many high-level events and meetings throughout the year, including the G20 meeting of agriculture ministers held in Buenos Aires, Argentina. The theme of the World Economic Forum Meeting, held in January in Davos, Switzerland, was *Creating a Shared Future in a Fractured World*, lending itself to only limited discussion of food security and agriculture. The 73rd session of the UN General Assembly in New York brought tuberculosis and noncommunicable diseases, which cause more than 70 percent of global deaths, to the fore of political discussion, but failed to garner collective support for combating them. It did secure education commitments, namely to support girls' primary and secondary education and the education of refugees.

The international development community also rallied support for SDG2 on Zero Hunger. The first Global Parliamentary Summit against Hunger and Malnutrition was held in Madrid in October and engaged more than 200 parliamentarians from countries around the world to foster cooperation for achieving SDG2.[15] The T20 (Think20), a network of research organizations and think tanks from the G20 countries, presented a communiqué to the president of Argentina, who served as the 2018 president of the G20, offering 20 policy recommendations for addressing the world's most pressing challenges, including food security, climate change, and inequality. Three of the recommendations homed in on designing sustainable food systems, investing in agricultural and food systems research and development, and reducing food loss and waste. In November, IFPRI and FAO organized a high-level conference on Accelerating the End of Hunger and Malnutrition that brought together more than 600 policymakers, researchers, program implementers, and representatives of civil society to share country experiences and identify technologies and innovations that can help to ensure the achievement of SDG2.

These initiatives were consistent with a continued shift on the part of the international community toward transforming the whole food system. The EAT-Lancet Commission focused on accelerating the transformation of food systems in order to achieve the Paris Agreement climate goals and the SDGs; it released policy recommendations in early 2019. A report from an international panel of experts, published in October 2018, also called for a focus on all aspects of the food system: the report found that poor diets pose a greater public health threat than malaria, tuberculosis, or measles, and called for food systems that can improve availability of and access to fresh, nutritious foods for all.[16] In October, the World Food Prize was awarded to Lawrence Haddad and David Nabarro for their work in elevating nutrition to the top of the global policy agenda, underlining the connection between food systems and nutrition outcomes.

REGIONAL AND NATIONAL AGRICULTURAL TRANSFORMATION

Some of the most notable regional policy developments of 2018 were in Africa. In January, leaders at the African Union Summit underscored their commitment to the Comprehensive Africa Agriculture Development Programme (CAADP) and the Malabo Declaration (which calls for ending hunger and halving poverty by 2025) by launching the African Agricultural Transformation Scorecard, a first-of-its-kind tool that tracks agricultural progress across 43 indicators. Of reporting countries, 20 out of 47 received an agricultural transformation score of above 3.94 out of 10, and are considered on track to fulfill their Malabo commitments by 2025, with Rwanda, Mali, Morocco, and Ethiopia leading the group. However, Africa as a whole—with a continentwide score of 3.6—is not on track to achieve the commitments. In another sign of political will to fulfill the CAADP and Malabo goals, as of August 2018, 19 African countries had made progress in preparing their Malabo-compliant National Agriculture Investment Plans.[17]

Some national-level policy shifts were initiated, particularly with the aim of fostering nutrition-driven agriculture. Ethiopia's National Nutrition-Sensitive Agriculture Strategy, launched in 2017, established a nutrition team under the state minister. The country approved a food and nutrition policy in November and began mainstreaming nutrition into several agricultural subsector strategies, including extension, horticulture, and postharvest strategies.[18] In South Asia, Bangladesh launched a stakeholder consultation on its 2013 National Agriculture Policy to strengthen the linkages among agriculture, nutrition, and gender by, for example, expanding national production of nutritious and safe foods and strengthening agricultural value chains.[19] Pakistan launched its Multi-Sectoral Nutrition Strategy 2018–25 to guide the country's efforts to increase the coverage and scale-up of equitable and high-quality nutrition services.[20]

A WAVE OF PROTECTIONISM

Trade protectionism was a dominant theme in bilateral relations in 2018, going hand-in-hand with the growing isolationism that marked many countries' political climates. Perhaps the most noteworthy development was the trade war between the United States and China. The United States imposed tariffs on US$250 billion of Chinese goods, and, as of late 2018, was threatening US$267 billion more. In a series of tit-for-tat moves, China set tariffs on US$110 billion of US goods.[21] The United States also levied tariffs on steel and aluminum from Canada, Mexico, and the European Union, and was met with retaliatory tariffs from Canada on US$12.6 billion of US imports.[22]

Trade protectionism bodes poorly for global trade and economic growth. Recent IFPRI research indicates that, in the event of a full-blown US-China trade war, China would face a 40 percent reduction in imports of US agricultural products, and the prices of critical goods would rise. Soybean prices, for example, would rise by nearly 6 percent and cotton prices by nearly 8 percent in China.[23] Trade wars may have the greatest impact on rural areas in emerging and developing countries, which depend heavily on sectors such as agriculture and food processing that are particularly affected by trade belligerence.[24] Developing countries fear being used as alternative markets for dumping excess tariff-hit products. But, on the positive side, their exports could become more competitive in China.[25] China's reduction of import tariffs on more than 400 agriculture-related products, for example, may have already benefited exporting countries in the Asia-Pacific Trade Agreement.

Some countries took the cue from these trade disruptions to pursue protectionism. India, for example, sought to restrict imports, such as of consumer electronics, as a means to encourage domestic manufacture of these goods.[26] In June, the prime minister of Malaysia cited the US trade moves in his call for developing countries to protect their markets.[27]

But 2018 also saw major pushback against trade protectionism. In March, following the US withdrawal from the Trans-Pacific Partnership a year earlier, the remaining 11 countries negotiated a new Comprehensive and Progressive Agreement for Trans-Pacific Partnership, which has been ratified thus far by 7 out of 11 signatory countries. The United States, Mexico, and Canada renegotiated the North American Free Trade

2018 FOOD POLICY TIMELINE

GROWING NEED FOR URGENT HUMANITARIAN ASSISTANCE

The 2018 *Global Report on Food Crises* reports 124 million people faced crisis-level or worse food insecurity in 2017, up from 104 million in 2016.

AFRICAN LEADERS ESTABLISH FREE TRADE AREA

44 African countries sign the African Continental Free Trade Agreement—potentially covering more than 1.2 billion people and over US$2 trillion in spending.

AFRICAN UNION LAUNCHES SCORECARD

The new Africa Agriculture Transformation Scorecard tracks progress toward achieving Malabo Declaration commitments on agricultural productivity and development—only 20 of 47 countries reporting are considered on track.

JUN

MAY

APR

MAR

FEB

JAN

ACCELERATING INNOVATION FOR SUSTAINABLE DEVELOPMENT

G7 Ministers endorse the "Whistler Principles" to guide and accelerate innovation for development impact, including supporting the poorest; investing in local innovators; and using rigorous data.

COUNTRIES STRUGGLE TO ACHIEVE THE SDGS

The *2018 Sustainable Development Goals Report* finds that a fast-changing climate, conflict, inequality, and persistent pockets of poverty and hunger challenge efforts to achieve the SDGs.

JUL

DECLARATION FOR A SUSTAINABLE FOOD FUTURE

G20 Agriculture Ministers commit to working together to ensure food security and healthy soils, promote transparent agricultural markets and trade, and combat food loss and waste and antimicrobial resistance.

EUROPEAN UNION REGULATES GENE-EDITED CROPS

The EU Court of Justice rules that gene-edited crops–including those using CRISPR-Cas9 technologies–are subject to the same stringent regulations as conventional genetically modified organisms.

AUG

SEP

GLOBAL HUNGER CONTINUES TO RISE

UN report finds the number of undernourished people increased to nearly 821 million in 2017, up from 804 million in 2016 and 784 million in 2015.

CHINA ANNOUNCES US$60 BILLION FOR AFRICA

Chinese president pledges $60 billion to Africa, including concessional loans, credit lines, and special funding for China-Africa development and imports from Africa.

INDIA CELEBRATES NATIONAL NUTRITION MONTH

India's Ministry of Women and Child Development leads activities countrywide highlighting key nutrition issues, including antenatal care, girls' education, healthy diets, marital age, and sanitation.

OCT

WORLD FOOD PRIZE RECOGNIZES IMPORTANCE OF NUTRITION

Lawrence Haddad and David Nabarro receive the 2018 prize for their global leadership in elevating the issue of maternal and child undernutrition in the food security and development dialogue.

LIMITING GLOBAL WARMING WILL BE EXTREMELY CHALLENGING

The Intergovernmental Panel on Climate Change warns that limiting global warming to 1.5°C above pre-industrial levels will require unprecedented changes in land and energy use, industry, buildings, transport, and cities.

UNITED STATES REAUTHORIZES GLOBAL FOOD SECURITY ACT

US president signs the Global Food Security Reauthorization Act extending Feed the Future, the government's global hunger and food security initiative, for five years.

NOV

GLOBAL EVENT ON ACCELERATING THE END OF HUNGER AND MALNUTRITION

IFPRI-FAO event convenes global leaders, experts, and stakeholders to highlight success stories, share lessons learned, and explore innovations to end hunger and malnutrition.

MALNUTRITION RATES UNACCEPTABLY HIGH

The *2018 Global Nutrition Report* advises that although the global burden of malnutrition is unacceptably high and affects every country in the world, ending malnutrition by 2030 is still possible.

DEC

PARIS AGREEMENT "RULEBOOK" ADOPTED AT COP24

Delegates at the UN climate meeting adopt rules to implement the Paris Agreement, but a lack of details leaves experts worried that the rules are insufficient.

FOOD SECURITY CRISIS IN YEMEN REMAINS DIRE

16 million Yemenis–more than half the country's population–face severe acute food insecurity despite ongoing humanitarian assistance, and their future remains insecure.

Agreement, resulting in a new deal called the United States-Mexico-Canada Agreement, set to go into effect when ratified by the countries, likely in 2020. Negotiations for the ASEAN-led Regional Comprehensive Economic Partnership (RCEP) were accelerated in an attempt to smooth the uncertainties resulting from the US-China trade dispute. The RCEP is poised to become the world's largest free trade bloc, covering 3.5 billion people.[28] In July, the European Union and Japan signed the Economic Partnership Agreement, which will create an open trade zone for more than 600 million people.[29] Also as of July, 49 African countries had signed the Continental Free Trade Agreement, which will cover 1.2 billion people and represent US$2.2 trillion.[30] As of early 2019, 18 countries had ratified the agreement; 22 are required for it to enter into effect. These developments are good news for boosting agriculture and nutrition. The *2018 Global Food Policy Report* noted that trade in inputs, commodities, and ideas can boost agricultural productivity and improve nutrition by enabling access to a diversified food basket.

SHAKY FUNDING FOR DEVELOPMENT

Throughout 2018, ongoing bilateral and multilateral funding commitments to the development agenda remained uneven, unclear, and hard to predict. In November, donors committed to investing an additional US$1 billion in the Global Financing Facility to improve women's, children's, and adolescent girls' health in poor countries, falling far short of the US$2 billion needed to expand the facility's work to 50 countries.[31] However, the promised funding will be leveraged to increase its impact by linking it to US$7.5 billion of World Bank funds.

The United States' step away from multilateralism had mixed impacts on foreign aid. In late 2017, the US Treasury Department announced that it would defund the Global Agriculture and Food Security Program (GAFSP), which was formed after the 2009 G20 summit; the United States had been the largest donor to the program, providing US$653 million in public and private investment to raise developing country farmers' incomes and increase food security.[32] Despite the withdrawal from GAFSP, the 2018 US spending bill actually increased the budget for international affairs by 4 percent, and in October, the US president signed the Global Food Security Act, reauthorizing the government's Feed the Future initiative for five years, and called on the executive branch to put an anti-hunger strategy in place.

One interesting ripple effect of 2018's rise in geopolitical tensions was in the area of foreign investment, to the benefit of countries in Africa south of the Sahara. In September, China announced US$60 billion in financial support to Africa in a package that includes credit lines, grants, and investment financing. The following month saw the establishment of the US International Development Finance Corporation, which will supersede and expand on the mandate of the existing Overseas Private Investment Corporation. With twice the lending capacity of its predecessor, the new agency will provide US$60 billion in loans and insurance to companies interested in operating in developing countries (and offer local-currency lending) and will incorporate several USAID (United States Agency for International Development) financing offices.

Global flows of foreign direct investment decreased by 23 percent in 2017, with only a very modest recovery expected for 2018. Flows to the least developed countries decreased by 17 percent, with different impacts by region. Flows to Africa continued to decline, decreasing 21 percent from 2016; flows to developing Asian countries remained stable; and flows to Latin America and the Caribbean increased by 8 percent. Foreign direct investment is the largest external source of finance for developing economies.[33]

Despite these developments, the private sector showed a burgeoning interest in using entrepreneurship to foster innovation, raise incomes, and slash poverty. The Africa Innovation Summit, for example, held in June in Rwanda, gave stakeholders a platform for finding action-driven solutions to pressing challenges such as energy, water, food insecurity, nutrition and health, and governance.[34] In October, the first-ever Nutrition Africa Investor Forum was held in Kenya, attracting business leaders, policymakers, donors, and development practitioners. The event gave dozens of small and medium enterprises the opportunity to vie for US$82 million worth of investment opportunities.[35] The 73rd UN General Assembly featured the launch of several tools to measure corporate progress on the SDGs as well as discussions on how innovative technologies, such as early warning systems, can be used to prevent famine and ensure food security.[36]

LITTLE ACTION ON CLIMATE CHANGE

2018 was a troubling year for climate change developments. A particularly alarming report by the Intergovernmental Panel on Climate Change, released in October, warned that the world is quickly reaching a point of "no return": limiting global warming to 1.5°C above pre-industrial levels will require rapid and unprecedented changes in many aspects of society, including the management of land, energy, cities, industry, and transportation.[37] A 20-year review released by the UN found that direct economic losses from climate-related disasters, mostly floods and storms, rose by 151 percent over the last two decades. Between 1998 and 2017, 4.4 billion people were injured or displaced by such disasters.[38] Climate shocks, mainly droughts, were a primary driver of food insecurity in 23 countries in 2017.[39] The outcomes of policy and regulatory processes for innovative agricultural technologies such as genetic engineering and synthetic biology, which could potentially boost agricultural productivity and resilience to climate change, remained uncertain in 2018. These types of regulatory decisions will have an impact on whether and how such technologies are adopted by low-income countries.

Commitments to meeting the Paris Agreement goals varied wildly against the backdrop of anti-science sentiment in some countries. The abandonment of the climate agreement was a consideration in the Brazilian presidential election, for example. The US administration stopped federal implementation of the agreement in 2017, but at the same time, 22 US states, 550 cities, and 900 companies with US operations have made climate commitments that could reduce US emissions to between 17 and 24 percent below 2005 levels and bring the country close to meeting its Paris Agreement targets by 2025.[40] The development of renewable energy from solar and wind boomed. The Powering Past Coal Alliance of more than 20 countries (including Germany, Europe's top coal user) was launched in late 2017 to phase out the use of coal. New Zealand's new government announced its ambition to reach net-zero carbon by 2050. Despite these developments, the sum of current country-level climate change policies will still lead to a global 3.1–3.5°C increase above pre-industrial levels by the year 2100.[41] At the 24th Conference of the Parties to the United Nations Framework Convention on Climate Change (COP24) held in December, countries agreed on many rules for operationalizing the Paris agreement, but stalled on establishing rules for voluntary market mechanisms.

PROTRACTED CRISES

2018 saw the continuation and, in some cases, escalation of conflicts that pose grave risks to food and nutrition security. Conflict and insecurity were the main drivers of food insecurity for 74 million people in 18 countries in 2017 (primarily in Africa and the Middle East), and the number of food-insecure people needing urgent humanitarian assistance increased 11 percent over 2016.[42]

Yemen, racked by a civil war exacerbated by the entry of a Saudi-led coalition into the conflict, confronted a serious humanitarian crisis. As of late 2018, 16 million Yemenis, or half the country's population, were on the brink of famine. Some 1.8 million Yemeni children were malnourished, leaving them more susceptible to diseases such as measles and cholera.[43]

The collapse of Venezuela's economy in 2014 led to a full-blown crisis marked by price inflation, dwindling supplies of food and other vital provisions, rampant unemployment and outmigration, and malnutrition. Statistics on malnutrition are unavailable due to the Venezuelan government's blackout of health data, but a leak of health bulletins from the Ministry of Health in April 2018 revealed a 30 percent increase in infant mortality in 2016.[44] Surveys carried out by universities indicated that Venezuelans lost an average of 11 kilograms in body weight in 2017.[45]

Years of civil war in South Sudan, combined with economic despair and the worst harvest since the country's independence in 2011, resulted in the largest refugee crisis on the African continent since the Rwandan genocide.[46] In January 2018, a time of year when food is usually most abundant, nearly half the country's population was food deficient. Hyperinflation has meant skyrocketing food prices and put the relative price of a meal in South Sudan among the highest in the world—South Sudanese must spend 155 percent of their daily income for a plate of bean stew.[47] However, September saw the signing of a peace agreement that may finally end the war. Ethiopia and Eritrea also signed a peace agreement in July, ending a stalemate that began in 2000.

LOOKING AHEAD TO 2019

By all estimates, 2019 may be another difficult year. Global economic growth is projected to slow over the next two years, decreasing to 2.9 percent annually in 2019 and 2020. A tightening of monetary policies in emerging and developing economies, combined with a loosening of fiscal policy in the United States, is likely to increase global interest rates and, as borrowing rates rise for poorer countries, may dampen investment. Setbacks or negligible growth in per capita GDP are projected for central, southern, and western Africa, the Middle East, and Latin America and the Caribbean.[48]

Heightened risks for global trade flows and the impacts of El Niño will increase uncertainty for the livelihoods and food security of both producers and consumers.[49] Threats to nutrition also abound. Stagnation in improving the nutritional status of children and women will impact future generations. The demand for dairy, sugar, and vegetable oils is expected to grow, the latter two reflecting a rising demand for processed foods.[50] World sugar production already reached record levels in 2017/18, posing a challenge for addressing booming rates of overweight and obesity that are driven by rising incomes and increased access to high-calorie, low-nutrient "convenience foods."

These worrying trends will undoubtedly affect rural areas most. Rural economies, already long neglected, continue to struggle. Any slowdown in national economic growth will impact employment levels and absorption of growing populations into the rural labor supply. In most developing countries, the majority of poor people live in rural areas and bear the brunt of higher mortality rates, food and nutrition insecurity, and lack of basic services. High levels of poverty, as well as dependence on rainfed agriculture in much of Africa south of the Sahara, also leave rural residents especially vulnerable to the impacts of climate change.

As this *Global Food Policy Report* finds, focusing on the needs of rural areas is one of the most practical ways to achieve the SDGs and address many of the roadblocks encountered in 2018, from climate change to conflict to political instability. Rural revitalization represents a systemic approach to addressing poverty and food and nutrition insecurity through its recognition of the intrinsic links among sectors that make up the entire food system. An increasing number of organizations and alliances are using a food systems approach to tackle these issues. Rural revitalization also represents an opportunity to leverage some of the successes of 2018, mainly an upsurge of interest in using entrepreneurship, new technologies, and public-private partnerships to solve development challenges. In just under a decade, rural areas could become the premiere hubs of innovation, not only in agriculture but also in manufacturing and services, providing a means for many rural residents to move out of poverty, malnutrition, and a low quality of life, and perhaps even to stem the flow of rural-urban migration. The potential is vast.

The challenges encountered in 2018 will continue in 2019. Achieving SDG2–Zero Hunger by 2030 will require strong political will and both public and private investment. Policymakers, researchers, program implementers, and civil society are already crossing sectoral boundaries to learn from past experiences and scale up successes, as they did at the recent IFPRI/FAO Global Event on Accelerating the End of Hunger and Malnutrition.[51] With perseverance, 2019 can become the year when the will to eliminate hunger and malnutrition finally gathers momentum, forging a bright future for poor people around the world.

"Rural revitalization is timely, achievable, and, most important, critical to ending hunger and malnutrition in just over a decade."

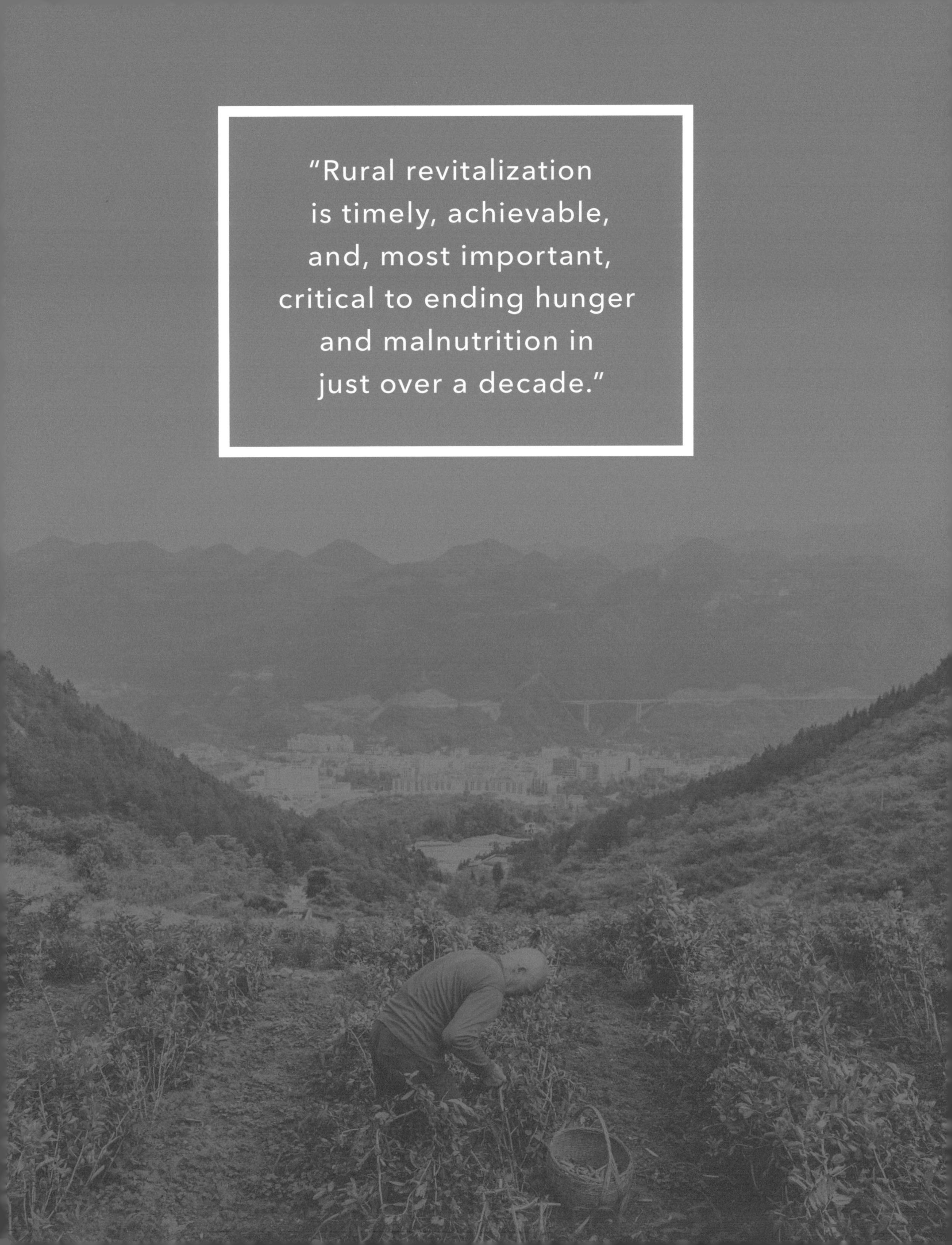

CHAPTER 2

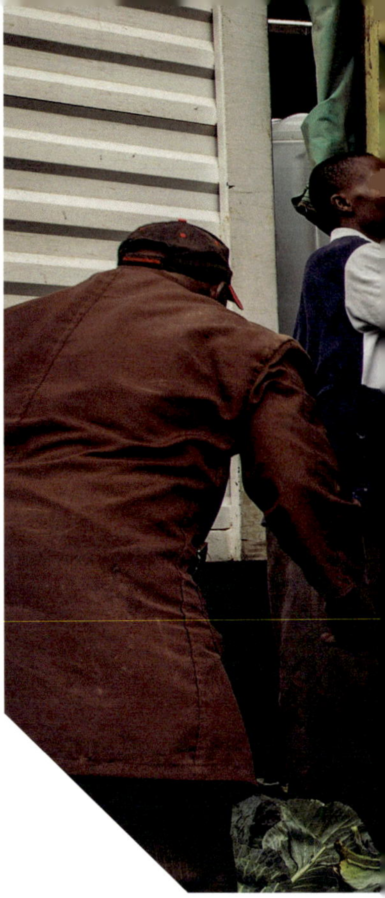

Rural Revitalization
Tapping into New Opportunities

ACHIM STEINER AND SHENGGEN FAN

Achim Steiner is administrator of the United Nations Development Programme, New York, USA, and **Shenggen Fan** is director general of the International Food Policy Research Institute, Washington, DC, USA.

KEY FINDINGS

- Rural revitalization requires a transformative approach that considers all aspects of making rural areas a good place to live and work for present and future generations.
- *Rurbanomics* is a development approach premised on the potential of symbiotic rural and urban systems to transform rural areas. Rural-urban links can spur growth and diversification in the agriculture and nonagriculture sectors, increase incomes, support value chain development, and improve well-being.
- Revitalizing the world's rural areas through a *rurbanomics* approach holds the key to achieving the Sustainable Development Goals, ensuring that everyone can contribute to and benefit from economic growth and development.

KEY RECOMMENDATIONS

- Adopt *rurbanomics* as an approach for strengthening rural-urban linkages to promote rural transformation. Strengthening rural-urban linkages, from farms to small towns to megacities, can benefit rural labor, production, distribution, markets, services, consumption, and environmental sustainability.
- Transform agrifood systems to benefit rural areas. Investments—in agricultural research and development, postharvest rural activities, and climate-smart and nutrition-sensitive innovations—have the potential to transform key components of the agrifood system, improve diets, strengthen rural economies, and improve rural livelihoods.
- Scale up rural nonfarm economic opportunities and build capacity for employment. Developing markets and creating clusters and special economic zones in rural areas can leverage economies of scale and increase nonfarm employment. Improving primary and secondary education and vocational training can support a productive rural labor force.

- Improve living conditions in rural areas. Stronger social safety nets, better access to basic services, and a healthier environment can expand opportunities for better livelihoods, protect the most vulnerable, and make rural areas attractive places to live and work.
- Reform rural governance to improve accountability and outcomes. Rural revitalization requires a public sector that is transparent, capable, and responsive to rural needs. Devolution to the local level has potential to support people-oriented governance when local governments have adequate capacity, can raise their own revenues, and mechanisms are in place that ensure accountability.

In a world where most poor people continue to live, work, learn, access services, and create communities in rural areas, rural revitalization holds the key to accelerating progress on the Sustainable Development Goals (SDGs). Today, rural populations account for 45.3 percent of the world's total population and at least 70 percent of the world's extremely poor.[1] These rural populations remain the world's most vulnerable and marginalized—and the rural context makes this persistent poverty nearly inevitable. Rural communities suffer from rapid population growth rates, inadequate job and enterprise creation, poor infrastructure, scarce financial services, and limited access to social services, and they bear the brunt of climate change impacts. Without innovative and holistic rural revitalization approaches that leverage new opportunities and address growing challenges, achieving food security for all by 2030 will be difficult—perhaps even impossible.

Revitalizing rural areas can tap into the potential of growing urban demand and a neglected rural labor force to stimulate broad-based economic growth in both rural and urban areas, and create a driving force for reducing poverty, ending hunger and malnutrition, and improving livelihoods. While addressing the growing challenges of urban food security and nutrition is vital, it is also essential and complementary to reshape agrifood systems to make them healthy, safe, and sustainable and to support rural revitalization. Public policies and investment to support increased agricultural productivity, diversification of agricultural production, and development of agrifood value chains can create better rural jobs and improve nutrition and well-being. For example, research from India and China has found that government investments in agricultural research, rural education, and rural roads are extremely effective in reducing poverty.[2] In Nepal, increased road density and quality have been shown to lead to improvements in nutrition outcomes for children.[3] Rural revitalization can also create decent living and working spaces for rural populations. Investment in high-quality educational and social services can make rural areas powerhouses for economic growth and better living spaces. Revitalizing rural areas can also provide solutions to rising rural-to-urban migration, increasingly congested cities, a rapidly growing population of unemployed youth, and worsening environmental degradation.

THE TIME IS RIGHT

Rural revitalization is a forward-looking strategy that leverages new economic opportunities and addresses growing challenges. While the 1970s and 1980s saw multisectoral rural reforms in Asia, such as South Korea's New Village initiative (Box 1), and in the 1990s the World Bank promoted community-driven rural development, the potential of rural areas has been neglected over the last decades. Now the time has come for rural "revitalization"—to make rural areas more productive, sustainable, healthy, and attractive places to live.

New market opportunities created by growing urban areas and new technologies make this the ideal moment to launch global, regional, and national rural revitalization strategies. Urbanization is proceeding rapidly—over half the world's population already lives in cities, and by 2050 two-thirds are expected to be urban residents—changing the outlook for rural development.[4] Together urbanization and population growth have created a booming demand for food and for more diversity of food. This growing demand is elevating the importance of agricultural growth and diversification and spurring the development of agrifood systems—from food production and processing to transportation to marketing to consumption. Rural revitalization is a complementary endeavor to urbanization—it can benefit both rural and urban areas. Increased agricultural productivity, greater diversity of production, and functioning rural markets linked to urban centers can feed the rural poor—most of whom are net food buyers—through increasing incomes, as well as feeding growing urban populations. These developments can make rural areas hubs of innovation. Advances in information and communication technologies (ICTs), for example, are opening up opportunities to access distant services, and new technologies for harnessing renewable resources, such as wind, solar, and biomass energy, are making electricity more affordable and accessible. These technological advances can create more attractive jobs and slow the tide of youth migration to urban areas.

Rural revitalization is also crucial for protecting both rural and urban environments. As the latest special report from the Intergovernmental Panel on Climate Change on the impact of climate change somberly notes, the time to reverse course and conserve our natural resources is now. The role of agriculture in deforestation, land degradation, and greenhouse gas emissions can no longer be ignored. Farming accounts for 70 percent of tropical deforestation; it also accounts for about a quarter of global greenhouse gas emissions. A "business as usual" scenario, which includes cascading impacts of growing demand for food on land, water, and air, will only worsen these impacts and make rural livelihoods riskier and less attractive. Rural revitalization can attack these maladies from various angles.

Rural revitalization must go beyond agriculture and agrifood systems. To achieve its broad goals of better rural lives, revitalization needs to simultaneously strengthen rural-urban linkages and leverage these links to create nonfarm rural jobs, improve rural governance, and ensure universal access to basic services and infrastructure such as water, sanitation, roads, healthcare, education, and a healthy environment. *Rurbanomics* offers an approach to development that focuses on promoting strong rural-urban linkages to help rural communities benefit from new markets and services.

BOX 1 South Korea's New Village Movement

From 1971 to 1979, South Korea implemented Saemaul Undong (New Village Movement), a community-based integrated rural development program that brought rural development to the forefront of the national political agenda. The program aimed to close the growing income and living-quality gaps between rural and rapidly developing urban areas. This program first focused on village-level self-help projects, then quickly expanded into a full array of investments in irrigation, agricultural inputs (especially hybrid seeds linked to the Green Revolution), electrification, transportation, and agricultural rice supports. Mothers' Clubs helped women initiate income-generating projects and participate in decisionmaking.

As a result of these and other efforts, farm household income increased fivefold from 1970 to 1979, reaching parity with urban households. But for all its achievements, Saemaul Undong did not successfully integrate rural and urban development at the local level, and so rural depopulation continued to worsen in subsequent decades. The lessons from this model could inform countries whose national leadership is committed to linking national programs with local mobilization.

See S.-Y. Park, "Saemaul Undong for the 21st Century," *Journal of International Development Cooperation* 2 (2008); M. Douglass, *The Saemaul Undong: South Korea's Rural Development Miracle in Historical Perspective*, Asia Research Institute Working Paper Series No. 197 (Singapore: Asia Research Institute, 2013); E. R. Reed, "Is Saemaul Undong a Model for Developing Countries Today?" (paper prepared for International Symposium in Commemoration of the 40th Anniversary of Saemaul Undong, Korea Saemaul Undong Center, Seongnam, Korea, 2010).

RURBANOMICS: TAPPING INTO NEW OPPORTUNITIES FOR RURAL REVITALIZATION

Rurbanomics is a comprehensive approach to fostering healthy and productive rural, peri-urban, and urban spaces that prioritizes linkages across rural and urban economies. It is built on the recognition that both urbanization and rural transformation are necessary to achieve inclusive and sustainable development. It frames rural economies as equal partners with urban economies, recognizing the potential of symbiotic rural–urban systems to transform rural areas. The *rurbanomics* approach emphasizes the vitality of rural economies not only as drivers of food security and rural well-being, but also as springboards for national, regional, and global value chains and as providers of quality environmental services for a sustainable world.

In this regard, *rurbanomics* focuses development policies on (1) ensuring greater attention to rural needs; (2) developing local assets and facilitating their integration into a nationally diversified and sustainable economic base; (3) leveraging local opportunities based on emerging national and global changes; and (4) empowering local communities and governments.[5] Recognizing that compartmentalized approaches are often ineffective, *rurbanomics* offers a multisectoral, cooperative, and networked vision of rural development. The value of rural assets, capacity, and potential for economic growth argues for strategic investments in rural areas and rural–urban partnerships to promote rural transformation.

To make rural revitalization a reality, *rurbanomics* explores how rural–urban linkages can empower rural populations to achieve inclusive and sustainable development. The interdependence of rural and urban economies can be used to accelerate the process of structural economic transformation—that is, a gradual economic shift from reliance on agriculture toward growth of the manufacturing and services sectors, which offer more productive employment and better incomes. Improving agricultural productivity is undeniably crucial to the growth of rural and national economies, and there is ample evidence that enhanced agricultural productivity is an important means of expanding off-farm agriculture-based activities.[6] Promoting this transformation requires creating markets where they are missing or weak, supporting economies of scale, and expanding and deepening rural banking, insurance, and financial systems.[7] Rural–urban links should not only be strengthened between strictly rural areas and megacities, but should also encompass small rural cities and towns that can form a key bridge from agricultural areas to markets and nonfarm jobs for rural producers. In fact, most urban dwellers reside in cities and towns of fewer than 500,000 people that, like nearby rural areas, are also in need of employment opportunities and social services.[8] Expansion of manufacturing and services sectors can benefit both rural and urban laborers and consumers across the rural–urban spectrum.

With strong rural-urban links, changing diets among urban populations provide an opportunity for rural transformation. Production of more nutritious and healthy foods to meet urban demand—such as high-value horticultural products and animal-sourced foods—can support value addition in the agrifood chain, including development of processing, transport, storage, wholesaling, and retail operations. Value chains can leverage the latest science and technology innovations to help revitalize rural economies and align them with future demand. Equally important, greater diversity of agricultural production and higher rural incomes can improve rural diets and food security, and these nutrition-sensitive value chains can provide small-scale farmers, youth, and women with better livelihood options that can have positive spillover effects on education, health, and incomes.

Above all, such a rural–urban relationship should foster innovation, research, and development in agrifood systems that will make rural settlements more attractive to all rural dwellers, but particularly youth. Engaging youth in productive rural activities will increase rural incomes and welfare and reduce out-migration. Creating these opportunities will require investments that increase agricultural productivity and will require addressing vulnerability to weather-related shocks, limited farmer capacities, distorted incentives, poorly functioning markets for inputs and outputs, and weak institutional capacities to manage the risk of food insecurity. Rural revitalization will also need to provide support to extremely marginalized and vulnerable groups (including children, women, and the elderly).[9]

Expanding the nonfarm sector is equally critical. If rural revitalization is to take root, rural livelihoods, which account for 38 percent of employment in low- and middle-income countries, must diversify. Farming households will need to either move up, in terms of raising productivity and profitability, or move out of agriculture altogether to nonfarm jobs that often have greater potential to move people out of poverty.[10] Not all urban growth creates such jobs—in Africa, in particular, urban growth is leading to rapid growth in services rather than manufacturing, which offers greater income potential.[11] Yet a boom in the nonfarm sector is already transforming rural opportunities in many places. In Pakistan, for example, one-half of new nonfarm jobs created between 2005 and 2014 were rural.[12] This shift has many potential benefits. Nonagricultural sectors are likely to have higher labor productivity and can engage the rural labor force, particularly rural youth. Research in Ghana has shown that households headed by rural youth are more likely to transition away from agriculture as a primary source of livelihood, and with promising results: these same households have lower poverty rates than their peers.[13] The nonfarm sector can also contribute to closing the gender wage gap: a study in Myanmar's dry zone, for example, found less gender disparity in nonfarm wages than in on-farm wages.[14] Nonfarm employment can also affect education positively: for example, in rural Ghana, mothers' nonfarm entrepreneurship was estimated to increase their children's rates of secondary school enrollment by more than 25 percent.[15]

Rurbanomics also frames rural economies as providers of quality ecosystems, such as fertile land, clean water, and low carbon emissions. Clean water and sanitation, for example, can be provided through "clustering"— that is, the practice of grouping rural enterprises by geography or sector to achieve economies of scale. In rural Senegal, water management was transferred to communities, in the form of clusters of villages operating and maintaining shared motorized wells, which served to meet the minimum water consumption needs of rural residents.[16] *Rurbanomics* can also help address environmental health along agrifood value chains by transforming food production and distribution channels and influencing consumer habits. Climate-smart and sustainability considerations can be included in all areas of the food system, from using drought-resistant seeds, diversifying crops, and adopting techniques that raise agricultural productivity and reduce deforestation, to partnering with urban centers to reduce the carbon footprint of food transportation and consumer waste.

TOWARD A HOLISTIC APPROACH TO RURAL REVITALIZATION

Turning rural communities into growth poles that can drive rural revitalization requires holistic and integrated approaches. Some initiatives that could spur rural revitalization and foster rural–urban links are described below.

INVESTING IN RURAL ECONOMIES: Strengthening and diversifying rural economies should be rooted in agriculture and beyond—and are necessary steps toward revitalizing rural spaces. The quality of life of rural residents depends on good governance, including adequate provision of public goods and infrastructure and strong investment in education, health, and social protection. Increased investment in roads, electricity, and communications in rural areas will better connect these areas to market opportunities and basic services. Linking rural and urban areas through good roads to transport not only high-value, perishable foods but also goods manufactured in rural areas is one example. These links can stimulate rural enterprises, whether related to agriculture or not, and thus increase household income, reduce poverty, and improve access to other services such as education and health. In Bangladesh, for example, improved rural roads reduced extreme poverty by 3 to 6 percent and boosted secondary-school enrollment for both boys and girls.[17] These services could also be coupled with income-generation activities. For example, setting up energy systems and associated infrastructure can equip rural zones with power for a range of economic activities; the cost of operating the systems can be paid for from the resulting profits.[18]

Necessary initiatives to propel rural economies include enhancing agricultural productivity (such as irrigation, improved seedlings, fertilizers, and postharvest management) and increasing agricultural diversity as well as developing village enterprises. While agricultural productivity sparks the growth of nonagricultural rural activities, additional investments in innovation, incubation centers, industrial parks, and special economic zones can help boost rural enterprise development. Incubation centers can provide shared infrastructure and training to start-up enterprises (and in the process, ensure that these enterprises follow best practices in terms of sustainability and production). Integrated industrial parks with modern infrastructure and services can attract private investment into rural areas. Special economic zones can spur whole industries by creating economic incentives, from duty-free imports and exports to tax breaks. The Chinese example of Township and Village Enterprises has shown the potential of promoting rural growth poles and corridors (Box 2).

BOX 2 China's dynamic rural revitalization efforts: a *rurbanomics* approach

China's overall economic reform started in rural areas. Agriculture grew at more than 7 percent per year from 1978 to 1984, lifting millions of people out of poverty and hunger. China's Township and Village Enterprises (TVE) helped to move 223 million people out of farming to nonfarming activities between 1978 and 2006, at which point the initiative had grown to 23 million enterprises, provided 119 million jobs, and produced 40 percent of China's exports. The reform was complemented by social safety nets and rural health services. Despite this giant stride in creating rural nonfarm jobs, rural infrastructure and public services were insufficient, and the rural–urban income gap widened. In addition, the rural environment was severely degraded.

To address this growing gap, China's annual Policy Document No. 1, released in February 2018, announced a new strategy for rural revitalization. The multidimensional strategy seeks to modernize China's rural areas and farm sector by 2035 by enhancing infrastructure, technology innovations, supply-side structural reform in agriculture, rural public services, and rural governance reforms. If fulfilled, the plan would signal a turning point for the 43 million poor rural Chinese and could also slow or reverse rural outmigration. A key reform for revitalization will be integration of agriculture and rural development strategies, policies, and investments, as seen in the reorganization of the Ministry of Agriculture as the Ministry of Agriculture and Rural Affairs.

See T. Saich, *Governance and Politics of China* (New York: Palgrave, 2001); J. C. Hernández, "Xi Jinping Vows No Poverty in China by 2020: That Could Be Hard," *New York Times*, October 31, 2017.

INVESTING IN ICTS: Investments in ICTs offer a promising means to address unequal access to market information and enable rural and urban markets to work synergistically. The possibilities are endless, from transmitting information on agricultural extension and market prices via cell phones to mobile banking.[19] Complementary investments in input markets and financial services will be required to ensure that markets function efficiently and equitably. Promoting small and medium rural enterprises that use new technologies could have a differential impact on youth, given their capacity to use ICTs.

INVESTING IN EDUCATION: Education is critical to providing rural people with knowledge and skills to improve their livelihoods. Investments include, first and foremost, improving primary and secondary education. Better education systems have a strong positive impact on many indicators of human development, including income, labor productivity, nutrition, health status, and family planning. Maternal education, for example, is a strong predictor of child stunting.[20] Literacy is a basic necessity for day-to-day functions and for using new technologies, such as reading an SMS on the day's market prices. The highest-yielding investments in education include increasing preschool enrollment (US$33 benefit for every dollar spent) and primary school enrollment (US$7 benefit) in Africa south of the Sahara, raising student test scores (US$4 benefit), and ensuring secondary school completion (US$4 benefit).[21]

Conventional educational systems, however, are insufficient to support *rurbanomics* and advance rural transformation. Rural entrepreneurs, particularly in Africa south of the Sahara, tend to engage in easy-to-enter activities such as sales and trade rather than professional services.[22] Start-up costs are one factor in their decision, but education also represents a major barrier to accessing other industries. Using education to catalyze rural revitalization calls for vocational and professional training initiatives, as well as entrepreneurial development tailored to local short- and long-term economic opportunities. Enhancing rural education systems and matching training to local needs and opportunities can ensure that educated youth do not leave for the city and can propel growth poles and corridors across rural areas.

SUPPORTING RURAL HEALTH: Investments in health services can serve as a linchpin to improving labor productivity, nutrition, and quality of life for generations to come. Rural residents currently have a lower life expectancy and poorer health status than their urban peers.[23] Creating rural health education systems that can train health professionals appropriately for rural practice has great potential. Training nurses to carry out procedures traditionally reserved for physicians, for example, may help communities where nurses are the only health professionals available.[24] Other promising models for improving rural healthcare include telemedicine and training community health volunteers on a large scale. For example, the community-based management of acute malnutrition approach, carried out in more than 60 countries, achieved a 90 percent recovery rate among patients.[25] The Safe Motherhood Programme offers a successful model for promoting rural maternal health in Africa (Box 3). In addition, direct nutrition interventions (such

BOX 3 The Safe Motherhood Programme in Nigeria

The Safe Motherhood Programme, also called the Abiye Project, in Nigeria's Ondo State shows that investment in rural health systems can save lives of pregnant mothers and their children. The Abiye Project gave each pregnant woman a mobile phone for toll-free, 24-hour access to her physician, and stationed tricycle ambulances in communities to provide first aid to expectant mothers. Before the program began in 2009, fewer than 100 pregnant women registered annually for public antenatal clinics. The number rose to 346 in 2009, 2,791 in 2010, and stood at 2,427 in 2012, with no lives lost. Given this success, the program was replicated by all local governments in Ondo State in 2013.

See A. Odusola, *Accelerating Progress on Maternal Health in Africa: Lessons from Emerging Policy and Institutional Innovations*, UNDP Working Paper No. 11 (New York: UNDP, 2013).

as promotion of breastfeeding or fortification) and indirect interventions (such as water and sanitation) for rural revitalization can drive advances in increasing dietary diversity, eliminating childhood stunting, and reducing poverty.

FOSTERING SUSTAINABILITY AND A HEALTHY ENVIRONMENT: Rural areas can contribute to rural and urban environmental quality through sustainable practices such as conservation agriculture (for example, integrating trees into annual cropping systems) and rainwater harvesting, or innovations like payment for ecosystem services, which provide community members with incentives to manage ecosystem resources and protect biodiversity.[26] Improvements in resource tenure systems and removal of distortionary subsidies can create better incentives for sustainable resource use. Adopting community-based management systems for water and forests can also offer opportunities for shared prosperity, environmental sustainability, and social cohesion. The Community-Based Integrated Natural Resources Management Project in Lake Tana, Ethiopia, is a good example.[27] The project encouraged rural people to invest in land improvement, stimulated off-farm employment opportunities, and empowered women and the rural poor; results included more than 32,000 hectares of degraded land restored for agriculture, forestry, and fisheries.[28]

ENCOURAGING DEVOLUTION AND PEOPLE-ORIENTED GOVERNANCE: Devolution of governance is vital to taking development to the people. A decentralized system of governance that is participatory, transparent, and accountable and balances fiscal powers with assigned functions is better positioned to respond to local contexts, needs, and aspirations. An integrated local governance system has many functions: it should facilitate democratic accountability; strengthen rule of law, ownership, and security to encourage private sector and community investment; build administrative capacity for policy management and service delivery; enhance fiscal resources; and support collection and availability of information to promote sustainable rural revitalization.[29] Evidence from Guatemala has shown that community participation in the design and implementation of rural services can have a positive impact on the level of accountability and responsiveness of local governments' provision of services, especially in municipalities with competitive elections.[30] However, when local governments do not have buy-in from the community, sufficient ability to raise their own funds, adequate capacity, or incentives for good performance—such as competitive elections—they may not provide the services needed for rural revitalization.

Governments also have responsibility for ensuring the well-being of marginalized populations. This means using well-researched program designs for addressing gender inequality, as well as providing safety nets (also known as social protection) to the vulnerable and extremely poor. Some of the most successful social protection interventions in the world have been undertaken in rural areas. For example, Mexico's Programa de Educación, Salud y Alimentación (PROGRESA), later renamed Prospera, focused exclusively on rural households at its onset and eventually scaled up to cover one-quarter of the country's population. It had a significant impact on children's growth and intake of micronutrients like iron, zinc, and vitamin A.[31] Other instruments, such as rural insurance schemes and rural pension schemes, have yielded positive results in a number of countries.

The initiatives described here highlight the role of government in making large-scale investments in areas where private actors and markets do not find it profitable to engage. Such actions require sufficient governance capacity in rural areas as well as effective coordination with national-level entities. Good governance is a prerequisite to rural revitalization, in the sense that functioning and responsive local governments act in the interest of the people and provide high-quality services. But good governance can also be a result of rural revitalization. As rural areas are reinvigorated, they will draw populations back in, leading to higher rural densities and greater political clout. A more even geographical distribution of more educated and wealthier people can empower local stakeholders to hold governments accountable and ensure sufficiently funded rural budgets. In a virtuous circle, good governance can, in turn, continue to make rural development "responsive and more relevant to local needs, aspirations, and lives."[32]

AN ACTION AGENDA FOR RURAL REVITALIZATION

To meet the needs of the poorest and most vulnerable by 2030, rural revitalization must begin now. Different regions will have different needs. In Africa, the focus will need to be on agricultural and rural development; in South Asia, diversifying the rural economy to expand rural employment and stem migration is likely to be a priority; and in China, improving the rural environment and living conditions to attract young people back to the countryside will be critical. A set of evidence-based and entirely achievable policies, programs, and research actions can advance rural revitalization in all these regions:

- **ADOPT *RURBANOMICS* AS AN APPROACH FOR STRENGTHENING RURAL–URBAN LINKAGES.** Building and strengthening a symbiotic relationship between rural and urban centers is key for improving agrifood system production, distribution, markets, services, and consumption and creating opportunities for rural labor. *Rurbanomics* offers an approach for engaging rural and urban areas—not only megacities but also small towns and cities—in development of nutrition-sensitive, environmentally sustainable, rural–urban value chains and the necessary infrastructure, markets, and skills. Effective integration of rural and urban economies, livelihoods, and services can benefit communities across the rural–urban spectrum.

- **TRANSFORM AGRIFOOD SYSTEMS TO BENEFIT RURAL AREAS.** Agriculture remains an engine of growth for many developing countries and the best means of accelerating expansion of off-farm agrifood activities, such as food processing and transport, that add value and generate income for rural residents. Agriculture must be seen as a business enterprise and as part of the agrifood value chain that feeds rural and urban areas. This requires diversifying and modernizing agriculture, including inputs, adoption of climate-smart agriculture, postharvest management, and financial support for nutrition-sensitive and sustainable food production. Public sector investments are needed in agricultural research, especially research focused on sustainable agriculture and agricultural technologies, and development of infrastructure to support agrifood systems. Governments should also support agriterritorial development tools, such as special economic zones, and agribusiness incubators for nascent businesses that can boost agriculture and associated activities in the agrifood system.[33]

- **SCALE UP RURAL NONFARM OPPORTUNITIES AND BUILD CAPACITY FOR EMPLOYMENT TO IMPROVE INCOMES AND LIVELIHOODS.** Investing in the nonfarm economy to develop markets and financial services where they are missing or weak can help individuals expand their businesses, provide greater returns to labor, and expand wage-earning opportunities, particularly for poor, female- and youth-headed households. Development of the nonfarm economy can also increase the availability of services in rural areas and deepen connections with urban areas.[34] Creation of "clusters," growth poles, corridors, and special economic zones in rural areas can leverage economies of scale to bridge market gaps. Investing in educational and vocational training with an eye to future trends and emerging industries in rural areas can create a productive rural labor force for generations to come. Research is needed to fill key knowledge gaps related to rural–urban value chains and the challenges facing the heterogeneous landscape of rural nonfarm activities.

- **IMPROVE LIVING CONDITIONS IN RURAL AREAS WITH STRONG SOCIAL SAFETY NETS, BETTER ACCESS TO BASIC SERVICES, AND A HEALTHIER ENVIRONMENT.** Rural revitalization requires improvement in rural living conditions; revitalization must go beyond markets to provide the basic services needed for better lives, particularly for the poorest and most vulnerable populations, and ensure a healthy environment. Multiple models exist of how to successfully provide water, sanitation, education, healthcare, and social safety nets in rural areas. Likewise, evidence on models for improving rural productivity while protecting or improving the use of natural resources is growing. Strengthening the links between rural and urban areas can facilitate the provision of basic services in rural areas and contribute to increased support for healthy environments and sustainable provision of environmental services, but sufficient protections and safeguards need to be put in place to guard against unsustainable demand.

- **REFORM RURAL GOVERNANCE TO IMPROVE ACCOUNTABILITY AND OUTCOMES.** Good rural governance is essential to ensure that the public sector, from local to national, is responsive to rural needs and can be held accountable for delivering policies, investments, and services that support better rural lives. Delivery of infrastructure, services, and an enabling environment for development of rural-urban linkages and greater rural opportunities depends on well-provisioned rural budgets, governance capacities, and transparency. Devolution of governance to the local level—when governments can raise adequate financial resources, have adequate capacity, and can be held accountable by constituents—can be an effective mechanism for ensuring people-oriented governance that will drive rural revitalization.

AN IMMENSE UNTAPPED OPPORTUNITY

The deadline for achieving the SDGs is looming. We have just over a decade to transform the day-to-day reality of billions of poor food- and nutrition-insecure people and protect the natural resources and landscapes upon which the planet depends. The most realistic way to achieve such a dramatic transformation is to target change to the world's rural areas. Revitalizing rural areas is a necessary strategy for inclusive and sustainable development and for ensuring that remote and hard-to-reach communities are not left behind.

Rurbanomics offers an approach for fostering rural revitalization and achieving many of the SDGs through symbiotic relationships between rural and urban labor forces, production, distribution, and consumption. Emerging lessons from countries like Brazil, China, India, and South Africa have shown that putting *rurbanomics* to work requires that policies and programs must be accompanied by investments in better governance, institutions, social participation, policy processes, and engagement of the private sector. Strong coordination–"whole-of-government" (central, state, and local) approaches that encompass multiple sectors (agriculture, infrastructure, education, and health) and have the support of a range of partners (state, market, and civil society)—will be critical for transforming rural-urban systems.

Rural revitalization represents an immense opportunity to build vibrant rural areas that can attract and retain employed, educated, and healthy rural residents. The time is opportune—the continued wave of urbanization in most countries has increased the urgency of closing both socioeconomic and quality-of-life gaps between urban and rural areas, while simultaneously tackling growing challenges such as climate change, natural resource scarcity, and youth unemployment. Addressing these challenges now is far more cost-effective than attempting to mitigate their effects in the future. But more importantly, rural revitalization fulfills a compact to begin to rectify global inequality and afford the world's most vulnerable people a chance to live healthy, fulfilling lives.

CHAPTER 3

Poverty, Hunger, and Malnutrition
Challenges and Breakthroughs for Rural Revitalization

HOMI KHARAS AND LORENZ NOE

Homi Kharas is the interim vice president and director and **Lorenz Noe** is a senior research analyst at the Global Economy and Development program of The Brookings Institution, Washington, DC, USA.

KEY FINDINGS

- The world is not yet on track to achieve the Sustainable Development Goals (SDGs) by 2030. Poverty and malnutrition rates are falling in many places, but not fast enough.
- Many of the world's poor and malnourished people live in rural areas and are dependent on the agriculture sector for food and livelihoods, so addressing rural development needs is central to achieving the SDGs.
- Breakthroughs are needed in both policy and financing to foster the agricultural and rural growth needed to reduce poverty and hunger and spark rural revitalization.
- Critical challenges to reducing rural poverty and malnutrition include limited international and domestic investment, the current international trade environment, and climate change.

KEY RECOMMENDATIONS

- Expand and improve data on rural poverty and malnutrition to provide evidence for sound policymaking. Establishing a working group to propose improvements, particularly in disaggregated rural and other subnational data, could be a first step.
- Adopt a systems approach to addressing rural needs, including integration of natural resource issues, as well as those of other sectors, with agricultural production and agrifood systems development. A global bioeconomy council could help support this approach.
- Increase financial and policy support for reducing rural poverty and malnutrition by strengthening national and international accountability mechanisms and commitment to rural revitalization.
- Create an attractive policy environment for private sector and blended public-private investment in rural development that could reduce rural poverty and malnutrition.

Reducing rural poverty, hunger, and malnutrition must be the keystone of any effort to improve rural lives. The first two Sustainable Development Goals call to "end poverty in all its forms everywhere" (SDG1) and to "end hunger, achieve food security and improved nutrition and promote sustainable agriculture" (SDG2) by 2030. These goals seem distant, especially as the world becomes more urbanized and political attention has shifted away from rural areas, where most of the world's poor live and malnutrition is most prevalent. Despite progress, current trends suggest that most countries are not on track to meet the SDGs. International and national policies in place today for food security and nutrition, along with associated funding and investment, are inadequate to create significant change in these trends. But solutions are at hand. Improving the lives of the rural poor will require sustainable improvements in agricultural productivity that can foster structural transformation, raise rural incomes, and improve rural food security and nutrition. Some effective tools are available to help achieve these aims and put the world on track for the SDGs.

CHARTING TRENDS IN RURAL POVERTY, HUNGER, AND MALNUTRITION

Are poverty and hunger falling in rural areas around the world, and at what speed? These simple questions are at the heart of any analysis of whether we can meet the SDGs by 2030. But data are limited, especially on rural areas, and trends in various metrics of poverty, hunger, and their correlates are moving at different speeds, leaving the answers unclear. We review here trends from the primary databases on poverty and food and nutrition security, which focus on national-level data and available rural data.

Rural poverty, defined as the number of people in rural areas living on under $1.90 per person per day, is estimated to have fallen by half between 2005 and 2017—with just under 500 million rural people now considered poor.[1] With this rate of decline, many countries appear to be on track to meet the poverty targets, particularly the headline SDG1 target to end extreme poverty. The global rural poverty rate is currently 17 percent, in contrast to an urban poverty rate of 7 percent. For the limited countries with disaggregated data available, rural poverty rates are higher than urban poverty rates (using the $1.90 per day poverty line). However, the gap between rural and urban poverty rates appears to be narrowing as overall poverty rates decline.[2]

The global rate of undernourishment or hunger (insufficient food intake to meet dietary energy requirements) saw a steady decline over recent decades, until it began stagnating or reversing in the past couple of years—falling from about 15 percent in 2005 to about 11 percent in 2016.[3] The prevalence of hunger is highest in Africa south of the Sahara (22.3 percent) and South Asia (15.1 percent).[4] In terms of absolute numbers of people, an estimated 821 million people were suffering from undernourishment in 2017, up slightly from 2016 and not much improved from 2000.[5]

Global progress in reducing child stunting (low height-for-age) has been substantial, with reductions in prevalence from 39 percent in 1990 to 22 percent in 2017.[6] But 151 million children under the age of five are classified as stunted, with limited progress since 2000.[7] Stunting is indicative of inadequate nutrition and poor health, and the interactions between the two. For rural areas specifically, the picture is less clear, due to the lack of systematic data. National-level demographic and health surveys show that, for countries with disaggregated data, rural areas lag behind urban areas in reducing stunting rates.[8] While most countries are making progress in both rural and urban areas, the rural-urban gap is narrowing in only a few countries. Similar overall trends and mixed progress in addressing rural-urban gaps can be seen for underweight (low weight-for-age) among children as well.[9] Review of a number of case studies that disaggregate rural and urban data reveals mixed results—rates of undernourishment, stunting, and other indicators of hunger are often higher in rural areas, but substantial variation arises in the size of the gap between rural and urban areas.[10]

Disaggregating the data on malnutrition by country and extrapolating recent malnutrition trends out to 2030 for those official indicators that are currently quantifiable and available for most countries—stunting, wasting (low weight-for-height), undernourishment, and child overweight—shows that for each of these indicators, less than half the needed progress to end hunger by 2030 is likely under a business-as-usual scenario (Figure 1). In the case of child overweight, current trends are moving in the wrong direction—overweight and obesity are increasing and expected to become more prevalent through 2030 (Box 1).[11]

Making matters worse, business-as-usual trends may represent overly optimistic scenarios for many countries, because they do not reflect growing

FIGURE 1 Projected progress toward meeting SDG targets by 2030, business-as-usual scenario

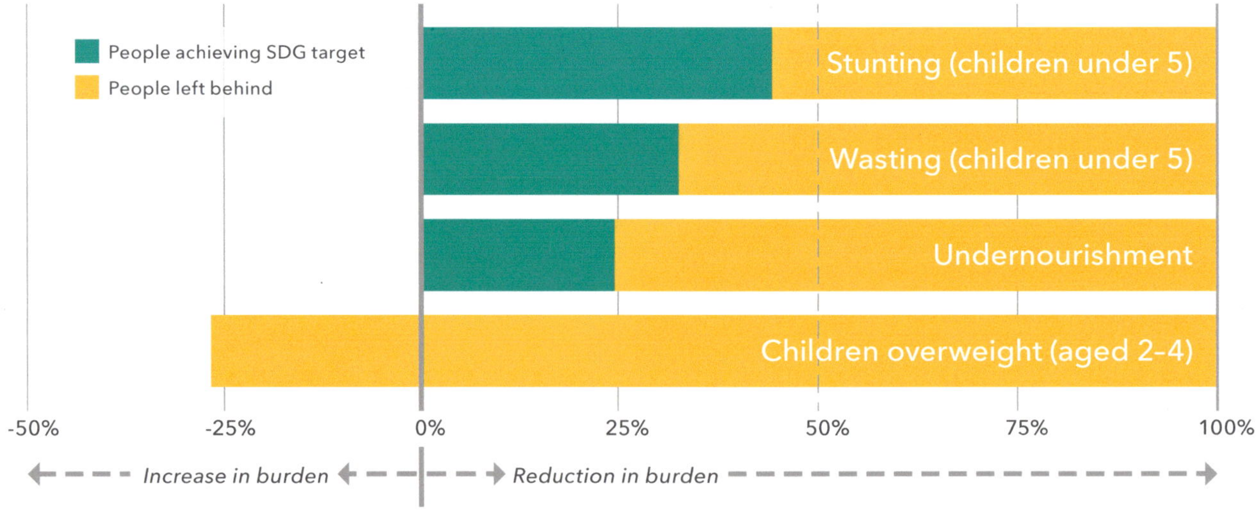

Source: H. Kharas, J. W. McArthur, and K. Rasmussen, *How Many People Will the World Leave Behind? Assessing Current Trajectories on the Sustainable Development Goals*, Global Economy and Development Working Paper 123 (Washington, DC: Brookings Institution, 2018).

BOX 1 Overweight and obesity in rural areas

The rapid global rise in overweight and obesity has been driven primarily by urban areas, but increasingly, rural populations in low- and middle-income countries are affected too, as incomes, livelihoods, and diets change. Recent data show that the overweight/obesity gap between rural and urban areas is shrinking, and in some developing countries overweight is now more prevalent in rural areas. For women, overweight and obesity are rising more quickly in rural than in urban areas in many countries. And a recent review found childhood overweight in 81 developing countries was only 1.08 times higher in urban areas on average than in rural areas.

See L. M. Jaacks, M. M. Slining, and B. M. Popkin, "Recent Trends in the Prevalence of Under- and Overweight among Adolescent Girls in Low- and Middle-Income Countries," *Pediatric Obesity* 10, no. 6 (2015): 428–435; G. A. Stevens et al., "National, Regional, and Global Trends in Adult Overweight and Obesity Prevalences," *Population Health Metrics* 10, no. 22 (2012); N. D. Ford, S. A. Patel, and K. M. Venkat Narayan, "Obesity in Low- and Middle-Income Countries: Burden, Drivers, and Emerging Challenges," *Annual Review of Public Health* 38 (2017): 145–164; R. E. Black et al., "Maternal and Child Undernutrition and Overweight in Low-Income and Middle-Income Countries," *Lancet* 382, no. 9890 (2013): 427–451.

challenges to food security and nutrition. Most notably, global warming could add 100 million more extremely poor people by 2030 and negatively impact food production and health.[12] Conflict also threatens progress; in many fragile and conflict-affected states, reductions in hunger are too slow. These challenges threaten to leave rural areas further behind.

Rural poverty and malnourishment are generally expected to move in the same direction, falling or rising together. For example, as a result of the food price spike in 2008, 100 million people likely fell back into extreme poverty, according to World Bank estimates, while hunger rose.[13] Because of these kinds of experiences, it is common for policymakers to assume there is a close correspondence between rural poverty and food insecurity and to address food insecurity through anti-poverty programs. For example, former chief economist of the Food and Agriculture Organization of the United Nations (FAO) Jomo K. Sundaram advocated strongly for social protection as the fastest route to ending hunger.[14]

In practice, we find a positive correlation between undernourishment and poverty at the global level, but the association is weak—just one-third of the movement in undernourishment is linked to movement in extreme rural poverty rates. Because rural poverty and undernourishment figures come from different sources and are determined quite differently, we should expect some divergence, but the degree of divergence between poverty and hunger measures highlights the need for better data.[15]

FOCUS ON RURAL AREAS

An estimated three-quarters of the world's poor live in rural areas, and most are among the 2.5 billion people who work in small-scale agriculture in developing countries.[16] Ending poverty and hunger for these rural residents will depend on well-functioning agriculture sectors and, more broadly, agrifood systems to help them lift themselves out of poverty and provide access to sufficient nutritious food.[17] Despite rapid urbanization, the transformation of rural areas offers the surest path to local livelihood improvements and food and nutrition security for the rural poor. Furthermore, revitalized rural areas can help drive healthy and sustainable agrifood systems in the context of urbanization and climate change. There is now ample evidence that, in the longer-term, GDP (gross domestic product) growth originating in agriculture is at least twice as effective in reducing poverty and undernourishment as GDP growth originating outside of agriculture.[18] This fact alone should convince policymakers to focus their efforts on agriculture and rural areas.

Growth in agriculture will depend in large part on getting effective technologies into the hands of farmers to raise agricultural productivity and on building food-system linkages with industries and markets. In Africa south of the Sahara, low agricultural yields are at the heart of persistent rural poverty, hunger, and malnutrition. The low yields are in part a problem of technology in its pure sense—unresearched and unimproved tropical crops have low yields and are not drought- and flood-resistant. But the technology challenge also lies in getting farmers to adopt modern practices. One approach is to foster markets for improved seeds, fertilizers, rural credit (and increasingly insurance), and farm produce, and technologies for reducing food loss and waste. In rural areas, most food loss occurs in the field or in storage and handling, requiring investments in postharvest handling and improved storage and transport. All these technologies need to be adapted at the local level—optimizing them for local meteorological and soil conditions, socioeconomic context, and the risks posed by climate change is as important as lab-based yield improvements. A complementary approach is to focus on the enabling environment, including rural infrastructure and legal and regulatory structures (roads, power, water, information, property rights). In both cases, public investments have been shown to be effective. For example, digital platforms that provide small farmers with information on weather and prices have had good returns in terms of productivity and incomes, as have research and development of seeds that are drought- and heat-tolerant, particularly when combined with improved water harvesting, integrated soil fertility management, and nitrogen-use efficiency.[19] Climate change will increase the importance of technology for adaptation and mitigation, including resilient infrastructure and climate-smart agriculture. But in many countries, funding for such public investments to increase agricultural productivity in the face of growing climate challenges is inadequate.

Increasingly, research and technology development are focused not just on farm yields but also on agribusiness and agrifood value chains to support rural transformation.[20] For development of these off-farm sectors, the question is what business models can provide incentives for the uptake of modern technologies and can be readily scaled. The answers will most likely come from the business community rather than from government, but can potentially be facilitated by the public sector through incentives, such as prizes and challenges, and through public-private partnerships.

INTERNATIONAL AND NATIONAL POLICIES AND FUNDING

Given the need to accelerate progress, policy and financing breakthroughs are critical to meeting the needs of rural areas. Although money is not the only answer to ending hunger and poverty, the amount of money provided by international donors and by national governments is a good indicator of the seriousness of and commitment to supporting rural development for achieving the SDGs. Public as well as private investments are needed to drive rural transformation. Policies are also crucial to success, including both international policies, particularly those governing trade in agricultural commodities, and national policies to support agricultural and rural development. Breakthroughs in all these areas are needed to reduce rural poverty and achieve SDG2.

INTERNATIONAL ASSISTANCE

Following the 2008 food price crisis, donor countries at the 2009 G8 summit in L'Aquila, Italy, declared their intention to invest in food security for low- and middle-income countries and pledged over US$22 billion in aid. The bulk of official food and nutrition security funding supports agriculture (59 percent) and rural development (16 percent) as well as food aid (17 percent, excluding emergency aid) and nutrition (5 percent)—all critical investments for rural transformation.[21] Both bilateral and multilateral institutions upped their gross disbursements in 2010, but in the following years, official finance for food and nutrition security plateaued (Figure 2).[22]

Overall, the share of food and nutrition security in total official financing has not risen over the last 20 years. A new financing mechanism established at L'Aquila, the Global Agriculture and Food Security Program, has a portfolio today (after almost 10 years) of only US$1.4 billion spread across 41 countries, and its largest donor, the United States, has not requested an appropriation for fiscal year 2019.[23] These amounts are nowhere near enough to move the needle on global outcomes. The L'Aquila pledges have apparently only served to sustain financing at historical levels.

In fact, the L'Aquila Food Security Initiative follows a historical pattern of short-lived increases in funding commitments in response to spikes in grain prices and little sustained commitment to addressing agricultural development and food security in developing countries. Though international assistance cannot be the sole source of support for improving food and nutrition security, if low- and middle-income country governments cannot rely on these funds, their ability and willingness to budget for rural food security-related programs will be weakened.

FIGURE 2 Funding for food and nutrition security, 2002–2016

Source: Authors' calculations, based on OECD QWIDS (https://stats.oecd.org/qwids/) and OECD Creditor Reporting System (CRS) (https://stats.oecd.org/index.aspx?DataSetCode=CRS1), accessed November 2018.

Note: FNS = food and nutrition security. Donor countries = participants in OECD's Development Assistance Committee (DAC). FNS gross disbursements and total assistance include official development assistance (ODA) and other official flows (OOF). FNS gross disbursements comprise OECD DAC categories: basic nutrition, agriculture, fishing, agro-industries, rural development, and food aid/food security programs.

INTERNATIONAL TRADE AND AGRICULTURAL POLICIES

Beyond direct assistance, the international community plays a key role in creating the conditions for reducing poverty and malnutrition through the trade and agricultural policies that shape developing country options. Ensuring food and nutrition security at the national level does not require that a country produce all its own food. Instead, most countries rely on agricultural trade to satisfy food requirements. However, global agricultural commodity markets remain distorted and thin, and therefore do not serve food security needs well. The nominal rate of protection (that is, the import tariff rate) for all agricultural products by all countries around the world has been increasing since the 2008 food price crisis.[24] Producer subsidies are also on the rise, with small declines in advanced countries more than offset by increases in subsidies in large emerging economies, such as China (although China has recently undertaken reforms to reduce the size of its domestic subsidies). Reduction or removal of such distortionary agricultural trade and macroeconomic policies could improve food and nutrition security. Open and accessible agricultural markets will become increasingly important to smooth food supplies, as climate change is expected to increase the variability in domestic food production in many developing countries.

Yet agriculture faces an uphill battle in international trade negotiations. Although the 2015 Nairobi World Trade Organization ministerial conference declared an end to export subsidies, the following ministerial conference in Buenos Aires in 2017 ended without an agreement on agriculture.[25] Ahead of the next ministerial conference in Astana in 2020, agriculture seems to have a weak hand in negotiations, constituting "only" 10 percent of global merchandise trade, while clearly being used as leverage to shape policy in other areas, as shown most recently in the trade confrontation between the United States and China over agricultural products.[26] Yet the fact that developing countries rely so heavily on agricultural products drives home the importance of trade as an asset in the global policy toolkit to improve poverty and food security outcomes.

For rural populations, the critical question is how trade rules and agricultural subsidies and tariffs affect the ability of and incentives for small- and medium-sized farmers to meet local food needs and to participate in larger markets. These farmers are likely to need support to identify and develop their comparative advantage in a market environment; for example, specialization in one commodity may allow them to take advantage of domestic or international trade to generate income, and then purchase food to generate dietary diversity.

DEVELOPING COUNTRY POLICIES

With donors seemingly reluctant to invest heavily in food and nutrition security, improving domestic spending and creating a better enabling environment for increases in agricultural productivity, market development, and adding value in food systems have become priorities for many developing countries. In this spirit, all African governments signed the 2014 Malabo Declaration, signaling their intent to deliver accelerated growth in agriculture to improve poverty and food security outcomes. Commitments include increasing domestic investment in agriculture, providing rural social assistance, boosting trade, and enhancing resilience to climate change. Three years on, 20 countries are on track in implementing the Malabo Declaration, meaning they have an average compliance of more than 3.9 out of 10 across the range of indicators that are used. Twenty-seven countries do not meet even this low bar, while the remaining eight countries have not reported on progress.[27] Only 3 out of 47 countries are on track to enhance investment in agriculture, and none spend at the target level of 10 percent of budget expenditures on agriculture.

Rural investment climates in countries in Africa south of the Sahara and other low- and middle-income countries have not improved since the mid-2000s, according to the Rural Sector Performance Assessments of the UN's International Fund for Agricultural Development (IFAD).[28] These measure ease of access to agricultural input and product markets and allocation and management of public resources for rural development, as well as accountability, transparency, and corruption. Particularly in low-income countries in Africa, where the need for rural development is greatest, the rural investment climate and rural corruption indicators have deteriorated in recent years. Some bright spots exist in the region, such as the improved policy and legal framework for rural organizations, increased access to water, and better dialogue between government and rural organizations. But given a decline in the overall institutional and policy climate, it is not surprising that rural development outcomes have not improved.

RURAL REVITALIZATION TO ACCELERATE REDUCTIONS IN RURAL POVERTY AND MALNUTRITION

Accelerating reductions in rural poverty and malnutrition will require investments for agricultural growth, including increasing the productivity of crops and agricultural labor, better access to markets, and stronger links with agrifood industries. Specific national and local context is important. However, looking ahead, there are some broad changes related to data and systems thinking, as well as commitment, accountability, and finance, that show promise for helping to address rural needs.

IMPROVING DATA

One source of the current stagnation is the weak feedback loop between outcomes and policy and institutional changes. As noted above, progress on ending rural poverty and hunger is difficult to track with the available data, and some important information, such as smallholder yields and amounts of food loss and waste, is not available for many developing countries. Understanding outcome trends at global, national, and, increasingly, subnational levels is a priority. Establishing a technical working group to propose ways of improving the quality and timeliness of outcome data—including disaggregated data by rural and urban areas—could be a first step toward improving key data.

The international community has already put forward several data initiatives. The recently agreed-to "50 x 2030" initiative from FAO, IFAD, the Global Partnership for Sustainable Development Data, and the Bill & Melinda Gates Foundation aims "to make improved agricultural data available in 35 countries by 2025 and in 50 countries by 2030," and will work to harmonize FAO and World Bank survey approaches to fill the information gap on developing-country smallholder farms.[29] At a broader level, participants at the UN Data Forum in October 2018 passed the "Dubai Declaration" in support of the Cape Town Global Action Plan for Sustainable Development Data, resolving to "ensure that quality, relevant, timely, open and disaggregated data at all levels of geography [...] are made available and accessible to all users" and strengthen the internationally agreed-upon framework for modernizing statistical systems.[30]

SYSTEMS THINKING

While technological changes and improvements will only be successful if they are taken up by small farmers, policy design must also recognize the interconnectedness of many issues across the food system that affect rural development, including the linkages through supply chains between farms and cities, and the linkages between food production and climate change and other environmental degradation. For example, a technological focus on increasing food production tends to ignore the sizable problem of food loss and waste, which may account for as much as one-third of total global food production. While there is some action on this front (for example, FAO's Save Food Initiative), far-reaching reforms remain difficult in part because of a lack of data on the scale of losses or the returns on policy actions.[31] Or, consider the negative repercussions of poor food safety, estimated to cost US$100 billion in lost productivity in low- and middle-income countries, with a disproportionate incidence of illness among poor and rural people.[32]

Consensus is growing that food and nutrition security should be treated as part of a broader system—an articulated systems approach can provide a framework for understanding how to increase agricultural productivity, reduce food loss, and improve food safety as part of a single system to improve rural incomes and food security. A stylized systems approach starts with the identification of needs defined as SDG targets, and then looks at the policies and investment resources required to achieve the targets.[33] Analyzing a country's needs allows policymakers to set priorities and identify linkages across poverty, undernourishment, smallholder yields, and exposure to shocks. Examining a country's policies allows for identification of gaps in safety nets, research, market incentives, regulations, and infrastructure. Finally, an accounting of a country's potential financial resources will establish the domestic and external financial commitments that can be marshaled to achieve the SDGs.[34] Data that can be shared among multiple constituencies are a prerequisite for any systems approach to function well.

Largely missing from the global policymaking landscape is an integrated vision of land, water, and energy use in agriculture—based on a systems understanding of interconnected issues—which is needed to address sustainability in a context of resource scarcity and climate change. As a result, important impacts and considerations are missed or neglected. For example, the fact that agricultural trade has yielded a 34 percent reduction in global water use is hardly ever cited as a benefit that should be taken into account in trade negotiations.[35] There is no "world water organization" to look at transboundary water management or "world land use organization" to review trade, research, and innovation in sustainable agriculture in a systemic way. Some efforts are underway, however. For example, the Food, Agriculture, Biodiversity, Land Use and Energy (FABLE) Consortium seeks to combine geospatial data on land use and agriculture to model national and global pathways for combating climate change that can address global objectives related to land use and food systems.[36] A global "bioeconomy council" could provide guidance and develop norms on this. Some individual countries, for example Germany, have tried to build these linkages under the auspices of a bioeconomy council, but there is no global equivalent.[37] The Rome-based food and agriculture agencies come the closest, but they are heavily oriented toward agriculture, although IFAD has taken an approach that emphasizes a more comprehensive development of rural spaces and of rural-urban linkages.

COMMITMENT, ACCOUNTABILITY, AND FINANCE

The international community has responded to the worrying trends in malnutrition and poverty in rural areas through a number of mechanisms, including pledges of funding for food and nutrition security, establishment of norms for sustainable agricultural investments, investments in agricultural and food research (primarily through CGIAR), elimination of global agricultural export subsidies, and creation of public food stockholdings.[38] But policy commitments have not translated into coherent and sustained finance for rural areas, and scaling up national and international public sector funding remains a challenge. Even where food and nutrition security projects can offer the most cost-effective solutions, in climate mitigation for example, they are not prioritized—funding for food and nutrition security hovers at around 8 percent of total climate mitigation aid. Taken along with continued obstacles faced by agriculture in trade and investment talks, the international commitment to addressing key needs of rural areas is weak and must be strengthened.

Given the limited global response, national and local governments must take greater responsibility. Indeed, most of the Voluntary National Reviews that developing countries submitted at the UN high-level political forum on Agenda 2030 report glowingly about their commitment to food and nutrition security.[39] Yet for all the outward show of political commitment, there is little discernible impact on outcomes and little accountability for the failure to follow through at either the international or national level. Compared to other development goals, food and nutrition security ranks toward the bottom among development leaders according to a recent global survey.[40] Both national and international political commitment must be reinforced by stronger accountability mechanisms.

PRIVATE SECTOR ENGAGEMENT

Engaging the private sector poses another challenge. The business case for rural investments that reduce poverty and improve food and nutrition security remains underdeveloped. Smallholder farmers are entrepreneurs operating in very difficult environments, where small scale and high risk are often prohibitive factors for private investment. Blended finance, that is, a combination of public and private funding, has potential to overcome these obstacles but has not yet taken off in the rural space. Development and strengthening of rural value chains, especially with foreign investment links, could also help, but can be expensive to organize. In addition to creating a supportive domestic and international policy environment for agricultural growth and providing essential infrastructure, private investment could be encouraged by new approaches that offer results-based incentives for the uptake of new technologies and could provide a platform for rural change.

A private sector window is included in the Global Agriculture and Food Security Program, but the results have been mixed. Since its inception in 2009, the private sector window has committed US$311 million in public funding, mobilizing perhaps five times that amount in private financing, and has reached about a million smallholder farmers around the world.[41] This falls well short of the scale of what is both needed and possible. The Business and Sustainable Development Commission estimates that there are profitable opportunities for US$320 billion of private sector investment every year in food and agriculture, which could provide jobs for 80 million people and unlock US$2.3 trillion in value.[42] However, identifying "bankable" projects in the area of sustainable land use is difficult; 30 percent of committed capital for sustainable land use remains undeployed because of the scarcity of appropriate projects.[43] One missing piece is a global platform to connect private, public, and philanthropic investors with profitable opportunities. Some efforts exist, such as the G20-led AgResults, which focuses on "pulling" mechanisms that crowd-in the private sector.[44] AgResults offers prizes to spur private sector research and development in areas such as low-emission production technologies and efficient on-farm storage. Globally, development finance institutions could do more to facilitate this matchmaking and to provide a broad suite of risk mitigation instruments and grants to offset the often small scale of appropriate private investments in the rural space.

Poverty, hunger, and malnutrition remain critical problems for many rural residents, and the challenges to improving rural opportunities are many. But breakthroughs in technologies, policy, and financing, supported by national and international commitments to small farmers and rural communities, could help to spark the rural change needed to meet the SDGs and to contribute to revitalizing rural areas.

"The transformation of rural areas offers the surest path to local livelihood improvements and food and nutrition security for the rural poor."

CHAPTER 4

Employment and Livelihoods
Connecting Africa's Rural and Urban Areas for Rural Revitalization

XINSHEN DIAO, PAUL DOROSH, MEKAMU KEDIR JEMAL, ADAM KENNEDY, AND JAMES THURLOW

Xinshen Diao is the deputy division director, **Paul Dorosh** is the division director, **Mekamu Kedir Jemal** is a research analyst, **Adam Kennedy** is a senior program manager, and **James Thurlow** is a senior research fellow in the Development Strategy and Governance Division, International Food Policy Research Institute, Washington, DC, USA.

KEY FINDINGS

- Many African countries have experienced strong economic growth and rapid urbanization, but not the industrial growth that has generated employment in other regions. Urban migrants working in the services sector rarely find jobs that are demonstrably better than those in rural areas or have much growth potential.
- Rural incomes in Africa south of the Sahara come predominately from agriculture (from crops, livestock, and wage labor). Policies that increase agricultural productivity and improve market access are vital to creating more and better employment opportunities in rural areas, where poverty is high and youth populations are large.
- Rapidly growing demand for food in urban areas offers promise for expansion of agro-processing and other agribusinesses. Growth in these sectors could support vibrant rural and small-town economies.
- Connectivity is improving, but many rural residents have little access to urban markets. About half of Africa's rural population lives far from urban centers.
- Integrating rural economies with small cities and towns shows promise for transforming rural Africa. Increasing rural–urban connectivity through roads and other infrastructure can drive rural growth and job creation.

KEY RECOMMENDATIONS

- Nest rural employment strategies, such as policies supporting agricultural commercialization and off-farm employment, within broader strategies for agricultural transformation and development.
- Modernize and diversify agriculture to promote youth employment, including adoption of modern technologies and development of high-value crops.
- Make complementary investments in basic services (electricity, communications) and human capital (education, healthcare) to promote vibrant rural areas and support better employment options.

Transformation and revitalization of the world's rural areas will only be possible if rural employment opportunities, both on-farm and off-farm, increase in number and offer better livelihoods. Rural employment in agriculture, agribusiness, and other rural-based businesses will be especially critical in Africa south of the Sahara, where the rural population continues to increase and where poverty will continue to be concentrated. Over the last two decades, rural livelihoods in Africa have diversified, with many household members now finding employment in nonfarm sectors, a process that mimics the transformation that took place in Asia and Latin America before the 1990s.[1] Migration from rural to urban areas to seek employment outside of agriculture has become an important livelihood strategy, particularly for better-educated young people. Nevertheless, in most African countries, rural livelihoods are dominated by agricultural income, be it from crops, livestock, or wage labor, making policies that increase agricultural productivity and improve market access vital to improving rural livelihoods.

As part of an employment strategy for rural areas, African countries will need to consider spatial aspects of rural economies, including investments in infrastructure and connectivity, such as transport networks and telecommunications. These connections reduce economic distance from markets, thus improving the profitability of economic activities (agricultural production and trade), and influence the flow of migrant labor. A nuanced public investment and employment policy that takes into account distinctions between large urban centers and outlying peri-rural areas (defined as those between one and three hours from a city of 50,000 or more people), and between subsistence and commercial farms, is needed. Investments in human capital that support income diversification and prepare rural youth for new rural and urban jobs will also be critical.

RURAL EMPLOYMENT, STRUCTURAL TRANSFORMATION, AND ECONOMIC GROWTH

Economic growth and development have historically been closely linked with structural transformation of economies—defined as increasing shares of both economic output (measured as value added, that is, gross domestic product [GDP]) and employment in the industry and services sectors, and corresponding declining shares in agriculture. Such a shift in output and employment typically results in higher incomes because average labor productivity is significantly higher in industry and services than in agriculture. Thus, a growth process that includes a shift in labor from lower-productivity agriculture to higher-productivity industry and services results in greater overall output and higher incomes.

Many African countries experienced an extraordinary period of economic development over the last two decades, driven by productivity growth within the agriculture sector and structural change.[2] As expected, this growth changed the composition of national economies, with agriculture shrinking as a share of both national GDP and the total labor force, and spurred rapid urbanization.[3] As of 2000, about 69.0 percent of the population of Africa south of the Sahara was rural, and 66.0 percent was employed in agriculture; by 2018, those percentages had shrunk to 60.5 and 57.0, respectively.[4] However, growth rates in industry, including manufacturing, have remained flat in much of Africa, while workers have moved into a burgeoning urban services sector. This pattern of growth has been characterized as "urbanization without industrialization."[5]

The African structural transformation based on services is problematic because much of the services sector comprises informal, labor-intensive activities, such as street vending, petty trading, and transport services, with labor productivity on par with traditional agriculture. These kinds of services do not offer a realistic pathway to creating the number and types of jobs needed to raise living standards for most Africans.[6] Creating more and better jobs is especially important given Africa's large youth population, estimated to make up 35 percent of the labor force in 2010, who will need to find remunerative employment.[7] While demand for manufacturing or nontraditional agricultural products such as fruits, vegetables, and horticulture products can be expanded either through export markets or by substituting domestic outputs for imports, services are mainly consumed in domestic markets. As a result, growth in the services sector is constrained by national demand that in turn depends on growth in national per capita incomes, population growth, and changing patterns of consumption associated with a shift to urban lifestyles. Thus, a services-led transformation will at best lead to modest rates of national economic growth.[8] Exploiting opportunities for promoting growth of productive jobs outside of services will be essential.

Transforming traditional agriculture is essential to generating better rural employment options and will be at the heart of revitalizing Africa's rural areas. In rural Africa, agriculture is still the dominant employer (accounting for over half of rural household incomes in most countries) and will likely be so for the foreseeable future (Figure 1). In contrast, in Bangladesh, as in most of Asia, nonfarm employment, along with government transfers, makes up the lion's share of rural household income. Africa's share of total world rural poverty is also expected to rise from 39.6 percent in 2015 to 58.1 percent in 2050.[9] But promising opportunities exist for increasing the productivity and profitability of agriculture. Demand for food is growing fast and projected to more than double by 2050, driven by population growth, rising incomes, rapid urbanization, a shift in diets toward greater consumption of higher-value fresh and processed foods, and more open intraregional trade policies.[10] Moreover, agricultural growth will help to leverage more rapid growth in the

FIGURE 1 Share of income-generating activities in total rural household income, by country

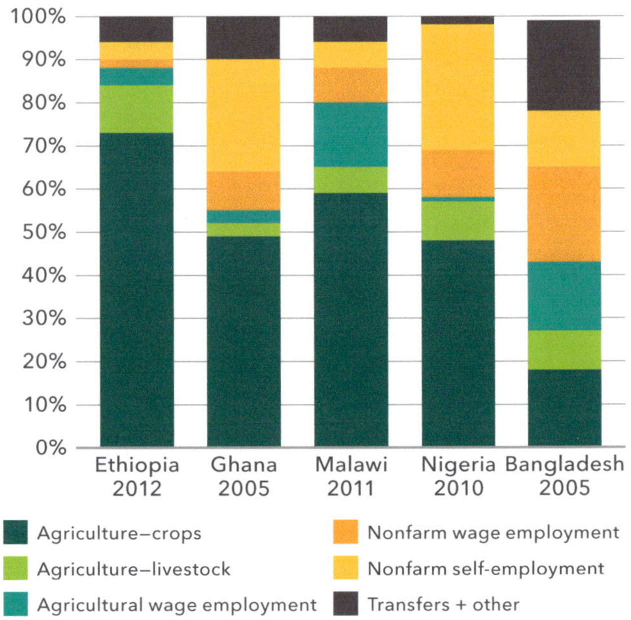

Source: B. Davis, S. Di Giuseppe, and A. Zezza, "Are African Households (Not) Leaving Agriculture? Patterns of Households' Income Sources in Rural Sub-Saharan Africa," *Food Policy* 67 (2017): 153–174.

larger agrifood system, including agro-processing and trading. According to a recent estimation, African agri-business (defined to include all aspects of the agrifood system except on-farm production) could become a US$1 trillion market by 2030, and that does not include the value of the additional agricultural production required to support growth in agribusiness.[11] Exploiting this potential will require a major shift from traditional to more intensive and market-driven farming practices, as well as the development of agricultural value chains and the broader agrifood system.

REDUCING ECONOMIC DISTANCE

Proximity to urban areas, including towns and cities, is a key driver of agricultural transformation. For rural areas near urban areas, access to urban markets provides higher agricultural incomes. This income can potentially generate enough demand for nonfood commodities to produce a vibrant rural and small-town nonagriculture sector, including grain milling and other agricultural processing, machinery repair, local transport, personal services (such as haircuts and tailoring), restaurants, and retail sales. Investment in public goods and connectivity was the foundation of the rural industrial development strategy in China and India in the 1970s and 1980s, and helped to spur the development of cottage industries to meet local demand for goods and services as agricultural incomes increased.[12] In Africa, the peri-rural areas are home to about 30 percent of the rural population and produce most agricultural output (Figure 2). In fact, as of 2015, about 80 percent of rural residents lived within three hours of an urban center, up from 57 percent in 2000.

However, about 20 percent of the rural population still resides in remote areas where agriculture completely dominates the local economy, the non-agriculture sectors are small, and little trade occurs with other regions. Farm incomes here are so low, and markets so limited, that any increases in food production are mainly consumed by the household. Limited markets also mean that increases in agricultural production are not likely to spark major growth linkages or a dynamic nonfarm sector.

FIGURE 2 Population of peri-urban, peri-rural, and remote rural areas in Africa south of the Sahara, 2015

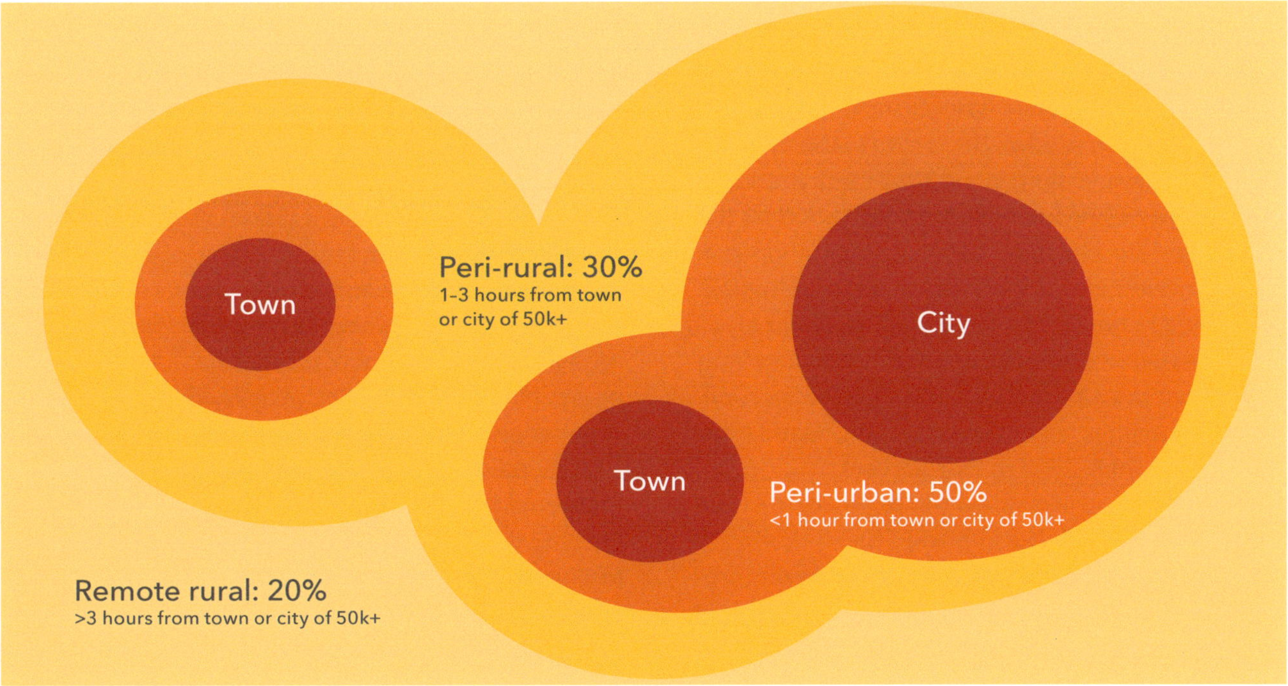

Source: P. Dorosh and J. Thurlow, "Agricultural Growth, Urbanization, and Poverty Reduction," in *Agricultural Development: New Perspectives in a Changing World*, ed. S. Fan and K. Otsuka (forthcoming).

A territorial approach designed to integrate rural economies with secondary cities and small towns shows promise for transforming Africa's rural areas. By putting stronger rural–urban linkages at the heart of rural development, a territorial approach can help to make diversification of rural livelihoods a viable alternative or complement to rural–urban migration.[13] Connectivity to urban centers with sizable markets is key to maximizing farm–nonfarm sectoral growth linkages, which have profound implications for agriculture and rural employment. Other boosts to rural businesses and employment stemming from urban proximity include greater access to services and credit, the availability of skilled labor, and investments in infrastructure and connectivity. Moreover, investments in Africa's agriculture sector are more effective than investments in urban areas in terms of reducing poverty. For example, economywide simulations conducted for Ethiopia and Uganda found that greater poverty reduction resulted from promoting agricultural growth than from poverty reduction strategies targeting urban areas, albeit at the cost of slower overall economic growth.[14] This relationship highlights the importance of promoting the rural nonfarm economy and its linkages with secondary cities and small towns.

Perhaps the most dramatic changes in connectivity have taken place in Ethiopia, a country with notably rugged terrain in its highlands, where most of the population resides. Beginning in the 1990s and greatly accelerating in the 2000s with the support of major development programs, including the Agricultural Development Led Industrialization strategy, the government of Ethiopia nearly quadrupled investment in rural areas, increasing the share of rural investment in total government spending from 8.5 percent to 23.8 percent.[15] The share of the rural population with access to electricity increased from just 0.4 percent in 2000 to 26.5 percent by 2016.[16] As a result of road investments and growth of urban centers, connectivity has improved dramatically. The share of the population living in remote areas more than five hours from a city of 50,000 or more people fell from 77.1 percent in 1984 to 21.6 percent in 2015 (Figure 3). Over the same period, the share of the population living *in* cities of 50,000 or more rose from 4.2 percent to 22.5 percent.

Substantial progress has been made in other parts of Africa as well (Figure 4). Rural access to electricity in Africa south of the Sahara continues to increase, with the share of the rural population with access more than doubling from 11.1 percent in 2000 to 24.8 percent in 2016.[17] Almost all of Nigeria's rural areas are now within three hours of a major city, and most areas are within one hour of a major city. Similarly, major investments in roads and bridges have significantly improved connectivity in much of eastern and southern Africa. In contrast, much of central Africa, including large parts of the Democratic Republic of the Congo, the Central African Republic, and Chad, remain remote. Creating remunerative employment opportunities in these areas of low population density will be extremely difficult without substantial investment in road networks. Moreover, areas that are far from seaports face similar difficulties in reaching world markets, though in countries with relatively high GDP growth, such as Rwanda and Burundi, farmers are able to sell to expanding national markets.

POLICIES FOR RURAL REVITALIZATION AND EMPLOYMENT

NEST RURAL EMPLOYMENT STRATEGIES IN BROADER STRATEGIES FOR AGRICULTURAL TRANSFORMATION AND DEVELOPMENT. To be most effective, a rural employment strategy should be consistent with a national development strategy that aligns urban, industrial, and agricultural policies, as well as public and private sector incentives and objectives. Identifying the best approach for specific locations requires looking at the important characteristics of the local economy, including its place along the rural–urban continuum and differences in agroecology that can determine the profitability of agricultural production.[18] Areas with an unsuitable agroecology for intensive production may experience little economic growth. Such areas need a balanced strategy that prepares youth for employment outside of agriculture while simultaneously creating the support systems necessary to build local resilience. In contrast, farms in peri-rural areas are generally more market-oriented and have greater scope for nonagricultural activities. Here, in addition to infrastructure to build connectivity, policies supporting agricultural commercialization and off-farm income diversification will be more important.

FIGURE 3 Travel times in Ethiopia: 1984 and 2015

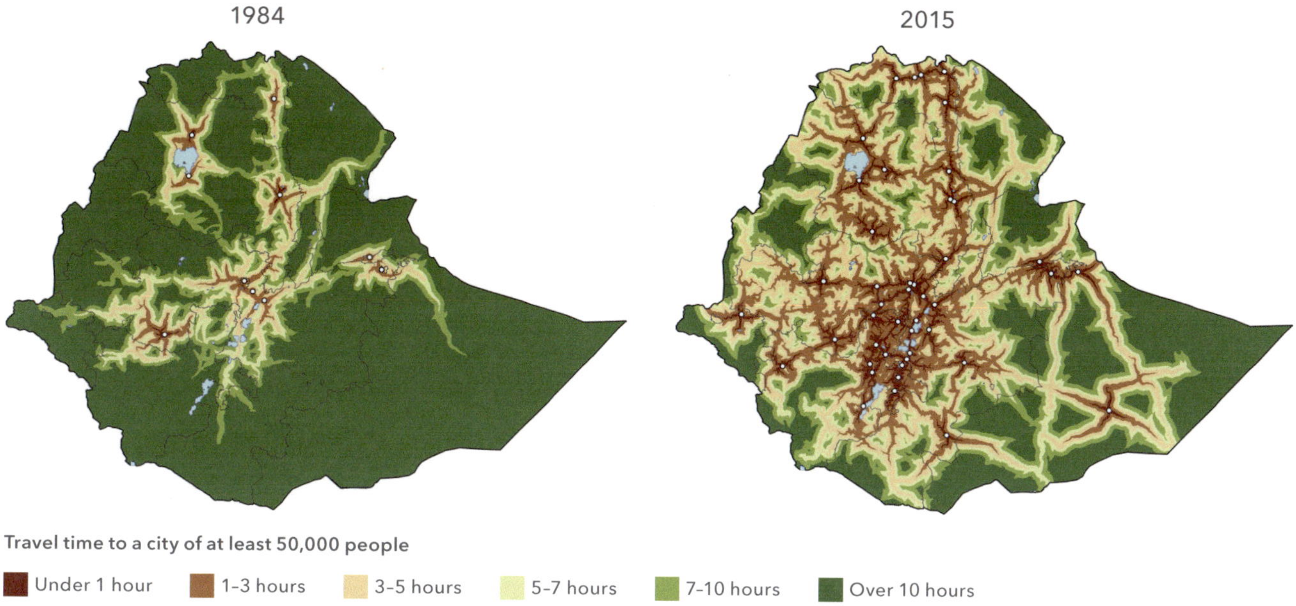

Source: E. Schmidt et al., *Ethiopia's Spatial and Structural Transformation: Public Policy and Drivers of Change*, Ethiopia Strategy Support Program Working Paper 119 (Addis Ababa: IFPRI, 2018).

FIGURE 4 Travel times in Africa: 2000 and 2015

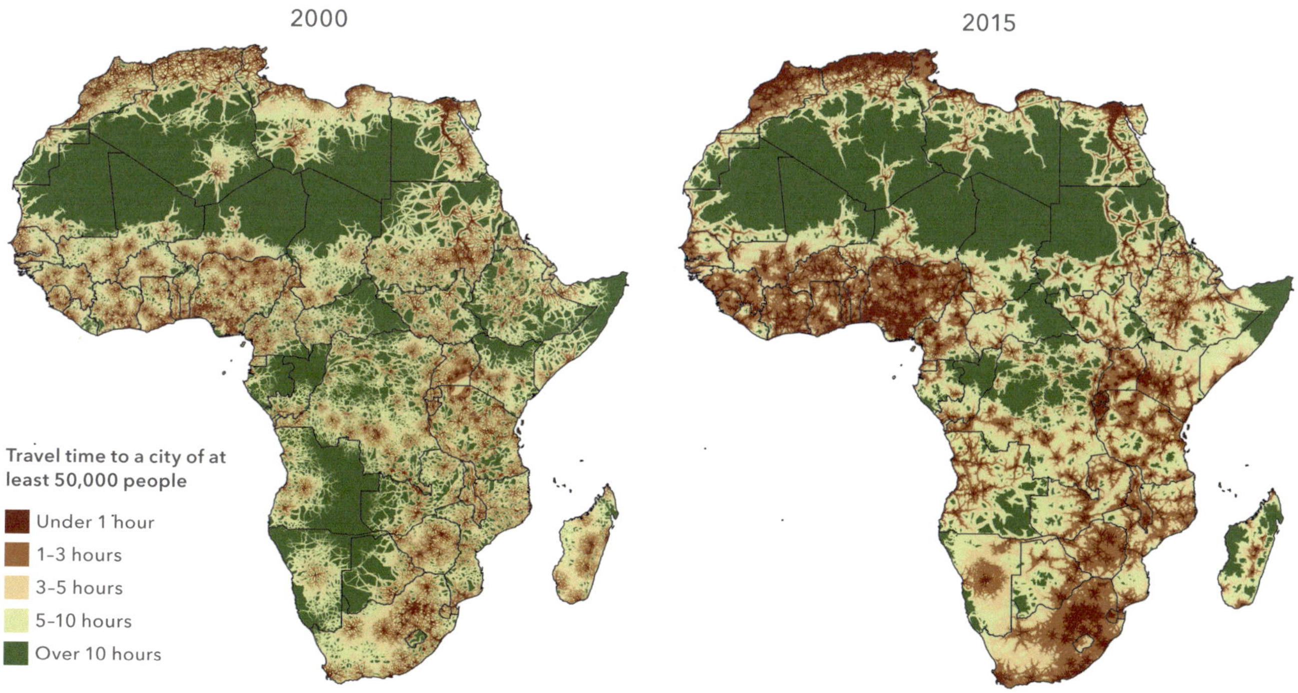

Source: Authors based on H. Uchida and A. Nelson, "Agglomeration Index: Towards A New Measure of Urban Concentration," background paper prepared for the *World Development Report 2009* (Washington, DC: World Bank, 2008) and D. J. Weiss et al., "A Global Map of Travel Time to Cities to Assess Inequalities in Accessibility in 2015," *Nature* 553, no. 7688 (2018): 333-336.

Note: As a result of improvements in the quality of data available, some roads may have been reclassified as lower quality in 2015, increasing travel times.

MODERNIZE AND DIVERSIFY AGRICULTURE TO PROMOTE YOUTH EMPLOYMENT. Making agriculture attractive to youth who aspire to urban-style livelihoods requires increasing agriculture's profitability, which depends on modern technology adoption, agricultural intensification, diversification into high-value crops, and commercialization. Mechanization services, improved seeds, and fertilizers (as well as the financial services to access them) are all technologies with potential to boost the productivity of traditional agriculture. However, these farm technologies are necessary but insufficient to achieve the agricultural growth and transformation needed for rural revitalization. Wider policies—such as trade and domestic price and marketing policies—can severely hamper structural change within the agriculture sector and discourage private sector investment. For example, in Malawi, huge investments in input subsidies to raise maize yields have not encouraged diversification into higher-value or more nutritious crops. And, as Bangladesh has learned, when governments promote specific crops such as rice, distortionary incentives and the generation of crop-specific technologies can hinder the diversification of agriculture—for example, into more nutritious pulses, vegetables, and livestock—and slow the process of agricultural transformation.

MAKE COMPLEMENTARY INVESTMENTS IN BASIC SERVICES AND HUMAN CAPITAL TO SPUR THE RURAL NONFARM ECONOMY. As rural youth become more educated and more rural households move from agriculture to the rural nonfarm economy in the near future, a much broader set of technologies than those discussed above will be required to make agriculture more productive and to make rural life attractive enough to keep youth in the countryside. Many nonagricultural policies and investments are complementary to farm inputs—such as electricity (including solar energy), telecommunications and Internet services (through cell towers or links to satellites), and social services (education and healthcare). These directly and indirectly improve agricultural performance and are especially important for high-value perishable products that growing cities increasingly demand. Moreover, such policies and investments are the foundation for a flourishing rural nonfarm economy to emerge as agricultural productivity and incomes increase and households consume more local goods.[19] Perhaps most important for long-term growth and poverty reduction, however, will be investments in human capital through primary and secondary education as well as technical schools.

BENEFITING FROM RURAL–URBAN CONNECTIVITY

Urbanization in Africa has not yet produced the type of structural transformation seen in Asia and Latin America, where the share of the labor force employed directly in agriculture has fallen sharply. Industrial development in Africa has been lackluster, and the services sector that employs many recent urban migrants affords little more opportunity than agricultural livelihoods in rural areas. The urbanization of poverty as more people move to cities is a real concern. But most evidence points to poverty being largely concentrated in rural areas through 2050. Redirecting scarce public spending away from the rural economy toward urban centers would risk worsening rural poverty by slowing growth in incomes and employment in rural areas. A more balanced strategy is needed that takes advantage of the rural–urban continuum and uses urbanization as a catalyst for economic development, starting with agricultural transformation.

Changing diets and growing food demand, particularly in urban areas, will open up new opportunities in the agrifood system and can spur significant employment growth in rural areas and small towns. Most of these opportunities will be realized by those living in peri-urban areas that are already more market-oriented and provide much of the food currently sold in urban centers. Agro-processing and trade have driven a large share of agricultural growth in Africa in the last decade and are likely to continue to expand with increased urban growth.[20] Investments in roads, electricity, and telecommunications infrastructure in peri-urban and rural areas will help foster rural employment growth through stronger linkages with small and large urban markets.

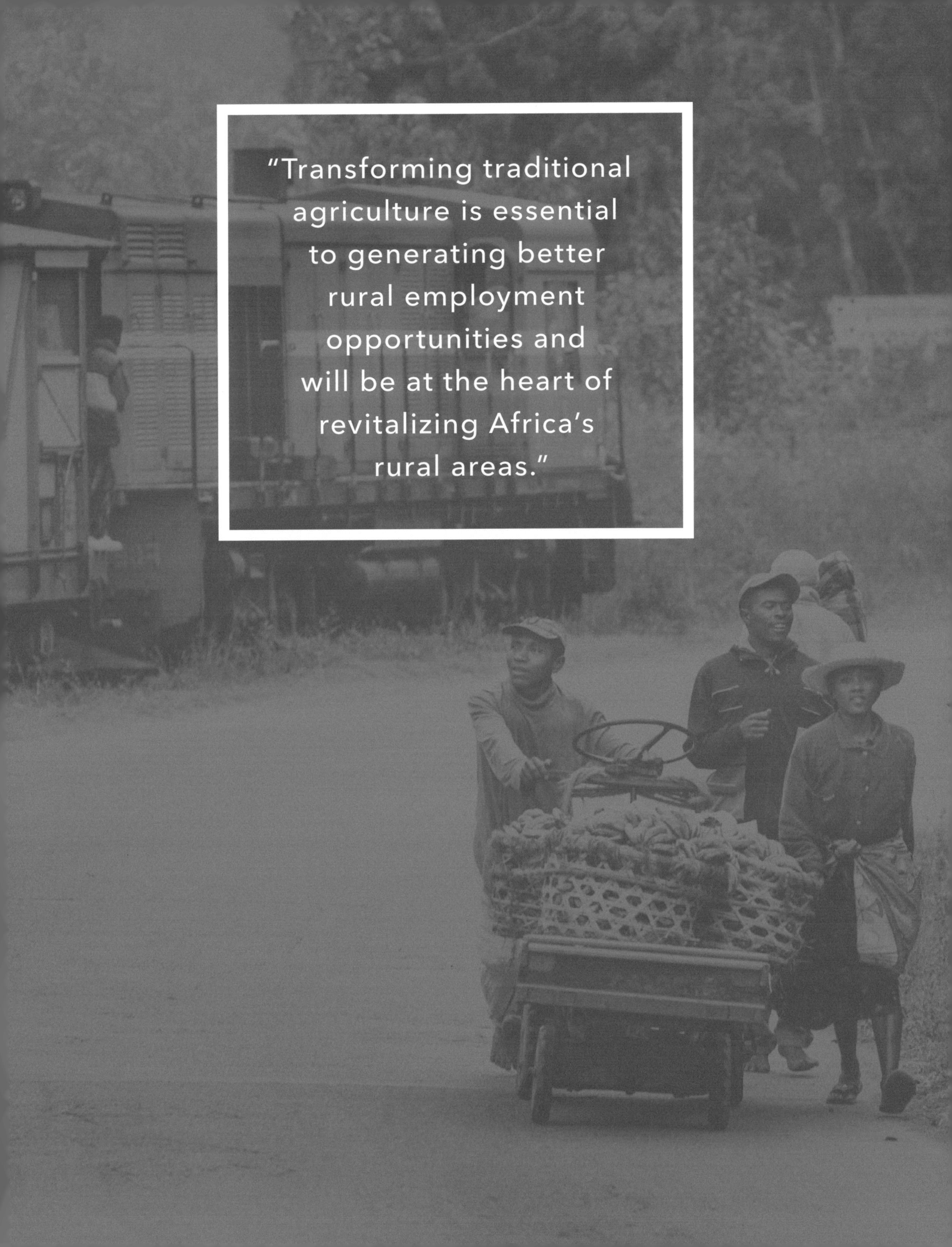

"Transforming traditional agriculture is essential to generating better rural employment opportunities and will be at the heart of revitalizing Africa's rural areas."

CHAPTER 5

Gender Equality

Women's Empowerment for Rural Revitalization

AGNES QUISUMBING, RUTH MEINZEN-DICK, AND HAZEL MALAPIT

Agnes Quisumbing is a senior research fellow, Poverty, Health, and Nutrition Division, **Ruth Meinzen-Dick** is a senior research fellow, Environment and Production Technology Division, and **Hazel Malapit** is a senior research coordinator, Poverty, Health, and Nutrition Division, International Food Policy Research Institute, Washington, DC, USA.

KEY FINDINGS

- Achieving gender equity and women's empowerment is of intrinsic value to women and girls and a key step toward achieving many Sustainable Development Goals and rural revitalization.
- Women and girls face a burden of time-consuming responsibilities, while controlling fewer resources and having less access to schooling, nonfarm jobs, and group membership, and less voice in governance and decisionmaking than men.
- Growth of nonagricultural jobs in many regions has led to the "feminization" of agriculture, with women taking on more farm-related responsibilities, often without greater resources.
- The *reach-benefit-empower* framework facilitates evaluation of how projects support women by ensuring their participation and that they receive benefits they value, and by strengthening their ability to make choices.
- Policies create conditions for successful projects, and should also be evaluated with the *reach-benefit-empower* lens. For example, policies to reduce women's domestic workload and to improve access to financial and agricultural services will facilitate empowerment.
- Policies to empower women can promote rural revitalization by supporting environmental sustainability, agricultural transformation, and development of nonfarm opportunities for men and women.

KEY RECOMMENDATIONS

- Use a *reach-benefit-empower* framework to ensure that interventions move beyond nominal participation to real improvements in women's lives.
- Increase women's effective voice through participation in formal governance structures, from local to national, and increasing women's confidence to become politically involved.
- Improve data and evidence relevant to increasing equality, particularly sex-disaggregated data and impact evaluation of programs and policies for women's empowerment.
- Include men and boys when designing projects and policies for women to reduce backlash and encourage changes in gender norms.

Rural revitalization requires that all members of society be able to participate in and benefit from growth and transformation of rural areas. But women and girls face particular challenges because of gender differences in resources and responsibilities. Achieving gender equality and women's empowerment is of intrinsic value to women and girls. It is also a key step toward achieving many of the Sustainable Development Goals, including eliminating poverty (SDG1), achieving zero hunger and malnutrition (SDG2), and ensuring good health and well-being for women and children (SDG3). Numerous studies have shown that when women control resources, their families, especially their children, enjoy better health and nutrition. Other studies have demonstrated that women's empowerment can improve agricultural productivity, dietary quality, and maternal and child nutrition.[1] These connections are especially evident in rural areas, reflecting the vital role that women play in agriculture, nutrition, and health.[2]

Despite SDG commitments to gender equality, women generally control fewer resources than their male counterparts and often have little say in how household income is used.[3] Women's "triple burden" of productive, domestic, and community responsibilities creates time constraints. The disadvantages that women face are passed on to girls, who often must help their mothers with domestic chores at the cost of their schooling. In societies where gender discrimination is entrenched, parents invest less in girls' health, nutrition, and education. Compared with young men of the same age, young rural women are more likely to be married, to have fewer years of schooling, and to own fewer assets. As a result, they are less likely to participate in the labor force or to obtain employment in high-wage sectors that are open to better-educated workers.[4]

The role of women and girls, and overcoming the constraints they face, is particularly important where men have migrated away from rural areas or moved out of agriculture. Historically, development has involved a process of structural transformation—a shift away from traditional agriculture as the manufacturing and services sectors become more important in the economy—that includes a reduction in employment in the agriculture sector as employment for both women and men expands in other sectors. Yet in many countries men move out of agriculture while women stay (or move out more slowly), and women's participation in agriculture consequently expands.[5]

This "feminization" of agriculture can create both opportunities and challenges for women. When remunerated, their increased involvement in agricultural work or in off-farm rural enterprises can empower women within their households and communities. However, if women are left with increased responsibilities in agriculture—without male labor, agricultural extension information, or recognized rights over land and agricultural assets, or without basic literacy and numeracy—they are unlikely to succeed. And if women continue to perform the bulk of unpaid work while men work in more lucrative jobs, gender gaps in wealth and labor burdens can widen.

How can the process of rural revitalization provide opportunities to empower women and girls? And how can empowering women and girls create sustainable rural communities? To answer these questions, we need to understand what we mean by women's empowerment. "Empowerment" has been defined as a process by which people expand their ability to make strategic life choices, particularly in contexts in which this ability had been denied to them.[6] The ability to exercise choice encompasses three dimensions: resources, agency, and achievements. *Resources* include not only access but also future claims to material, human, and social resources. *Agency* is the capacity to act independently and to make choices, including processes of decisionmaking, negotiation, and even deception and manipulation. *Achievements* are well-being outcomes, such as good health and nutrition. Rural revitalization must therefore include opportunities for women and girls to increase access and rights to resources (land, water, technology, and other assets), increase their decisionmaking power, and increase their well-being. Education and group membership can contribute by building women's confidence, social connections, and access to information.

To be sustainable, women's empowerment requires changes in gender norms and in deep-seated attitudes that affect relationships between men and women. This means that any programming involving women and girls must also consider potential effects on men and on families. Norms do not change in isolation—attitudes of the community must also change. To be successful, programs must facilitate changes in the attitudes of not only family members but also community members.

THE CHALLENGES

Rural economic transformations are bringing about rapid changes in rural demographic structures. As rural economies transform by increasing agricultural value added per worker, fewer workers are needed on the farm and labor can move from agriculture to nonagricultural enterprises. The reallocation of labor from agriculture to nonagriculture may involve migration, especially by young men. These changes have led to the feminization of agriculture in some regions, with an increasing share of agricultural work done by women, as noted above. A recent study argues that women's growing importance in agriculture in the Near East and North Africa, Central Asia, South Asia, and Latin America reflects a conjunction of factors: the diversification out of family farming induced by population growth and land fragmentation, the intensification of agricultural production (which can increase the need for female labor), the parallel growth of nonagricultural jobs, and changes in social and cultural norms that affect women's and men's mobility and livelihoods.[7] In Africa south of the Sahara, women already make up a large proportion of the agricultural labor force; and even in Latin America, where agriculture is regarded as a man's occupation, women's involvement is increasing through wage work in nontraditional agricultural export enterprises.

Even as women's involvement in agriculture increases, rural women continue to face constraints. Analysis of empowerment gaps in 13 countries shows, for example, that the biggest contributors to women's disempowerment are lack of access and power to make decisions about credit, excessive workload, and lack of group membership.[8] Women are less likely to own land or livestock, adopt new technologies, use credit or other financial services, or receive education or agricultural extension services—all of which are important for agriculture-based livelihoods.[9] Women also face significant mobility constraints relative to men that restrict their participation in markets and groups and their access to health and other services.[10] Constraints on women's time resulting from their larger agricultural responsibilities are compounded in many rural areas by the time needed for fetching water and fuel for domestic use. Lack of clean water and sanitation also leads to increased diarrheal diseases that affect women's health directly and add to their caregiving burden when other household members fall ill. Indoor cooking using traditional methods and low-quality fuels, especially when ventilation is poor, increases the prevalence and severity of respiratory infections.[11]

Governance structures may exacerbate gender gaps. Often governance structures—like farmers' associations, forest-user groups, or local government bodies—are dominated by men and do not take account of gender differences in needs and priorities, particularly for rural services such as agricultural extension and water supply. Even where women are represented, cultural norms or lack of education may prevent poor women from speaking in public, so their voices are not heard. This applies not only to local, regional, and national government representation but also to rural producer organizations, which are important providers of agricultural inputs and services.[12] In addition, poor governance of natural resources has led to environmental degradation. While this affects all rural residents, women and girls are particularly affected by degradation of water and forest resources because of their responsibilities for obtaining domestic water and fuel. Water-user associations, forest-user groups, and fishers' associations can play an important role in ensuring sustainable resource governance, but these organizations are also often dominated by men. Yet active involvement of women in the membership and leadership of forestry and fishers' organizations has been shown to improve natural resource governance and conservation outcomes, such as through broadening participation in rule setting and enforcement related to resource use.[13] Women's active participation in natural resource management groups, either through all-women's groups or strong involvement of women in mixed groups, is associated with greater collaboration, solidarity, reciprocity, and conflict resolution.[14]

Another constraint that impedes efforts to close the gender gap is lack of data and knowledge on the linkages between gender equality and rural revitalization—in particular, a lack of sex-disaggregated data on key outcomes. Although data on labor force participation and wages are typically collected at an individual level, which allows comparisons between genders, information on women's and men's ownership of assets, agricultural enterprises, and the like is rarely available. Employment data disaggregated by sex and age are not consistently available—national-level demographic and health survey data, for example, generally focus on women of reproductive age, and may not include the men in respondents' households. Women's participation in agriculture is also underreported because of perceptions that women do not farm or fish, even when they are active in many stages of production and processing. Data gaps may also result from poor understanding of the linkages among gender equality, women's empowerment, and rural transformation, and how these linkages may vary across the life-cycle. Finally, there are gaps in research on what programs and policies are effective in enhancing gender equality and empowering women in rural change.

REACH, BENEFIT, AND EMPOWER

A new *reach-benefit-empower* framework for assessing whether and how specific agricultural development investments empower women could serve to guide the design and implementation of programs and policies intended to stimulate rural revitalization by ensuring they promote women's empowerment. Originally developed to assess projects' gender approaches based on whether projects are designed to reach, benefit, or empower women, the framework can be expanded to assess how policies can empower women. The framework reflects how thinking about gender integration has evolved over time and draws on recent empirical evidence from past projects that integrated gender but often fell short in terms of benefiting and empowering women.[15]

REACH, BENEFIT, EMPOWER FOR PROGRAMS

A strategy focused on *reaching* women emphasizes engaging women in project activities and tracks progress in terms of participation—for example, measuring the number of women who attend meetings or receive training, the percentage of women in groups formed or supported by the project, or the percentage of women among those with access to extension or other services provided by the project. Some projects struggle from the outset with reaching women when they design groups, meetings, or trainings with men's needs in mind, choosing locations, times of day, and settings that are not culturally appropriate for women to attend or feasible given their schedules and workload. To ensure that women participate, projects often try to identify and alleviate gender-based constraints to participation—such as changing the time or place of meetings, forming women-only groups, or hiring women staff in the implementing organization or as lead farmers or extension agents. Projects may also require that women make up a certain share of the participants (typically 30 percent).

While reaching women is a basic measure of whether a project offers gender-equitable opportunities for participation, the ability of men and women to benefit from these opportunities is not equivalent. If a project strategy aims to *benefit* women and girls, the project design, implementation, and evaluation should focus on ensuring that whatever outcomes the project is seeking—reduced hunger, increased income, greater resilience—go to them. This requires going beyond *reaching* women and girls to ensuring that the project delivers *benefits* that women and girls value. General programs to revitalize rural areas, such as improved roads or electricity, can benefit women and girls; but their particular needs and constraints must be addressed specifically to ensure that they do benefit from these investments. Because women and girls are often responsible for domestic work, provision of clean domestic water supplies and clean cooking fuels is particularly important. Sanitation programs, especially the provision of latrines or toilets in homes and safe, gendered latrines in schools, are helpful in a number of ways, from reducing the burden of caring for the sick to encouraging girls' schooling.

If a project strategy aims to empower women and girls, then it must go beyond reaching or benefiting them. *Empowering* women and girls means strengthening their ability to make strategic life choices and to put those choices into action—through control over resources, participation in decisionmaking, or individual agency in domains that women themselves value. Such empowerment is key to sustainable and inclusive rural revitalization. However, empowering women and girls does not mean focusing exclusively on them and ignoring men and boys. On the contrary, projects focused exclusively on women may fail to consider appropriate roles and benefits for men, thus risking backlash in the household or community. This not only endangers the sustainability of the project, but also misses an opportunity for changing gender norms.

Figure 1 illustrates the differences between reach, benefit, and empower for a nutrition-sensitive agricultural extension project. Reaching women would only require that women receive a one-way flow of information, which might be measured by the number of extension agent visits to women, the number of women attending extension meetings or field days, or the proportion of women who have heard a radio program. Benefiting would require that the content of the extension information be

FIGURE 1 Applying the reach-benefit-empower framework for a nutrition-sensitive agricultural extension program

	REACH	BENEFIT	EMPOWER
OBJECTIVE	• Deliver agricultural extension services to women	• Increase women's well-being	• Increase women's agency over production and nutrition decisions
STRATEGIES	• Provide transportation • Conduct training during convenient times of the day	• Consider women's preferences and constraints in design and content of training	• Enhance women's decisionmaking power in households and communities, especially on which crops to grow
INDICATORS	• Proportion of women attending training, receiving extension advice	• Sex-disaggregated data for land use, yields, income, consumption, nutrition, time use, etc.	• Decisionmaking power over production, income, food consumption • Reduction of gender-based violence, time burden

Source: Authors.

useful to women, for example, that it be applicable to the crops or livestock for which women are responsible. Going beyond a one-way flow of information, participatory approaches such as farmer field schools that use group-based experiential learning have been shown to benefit women farmers though increases in crop and livestock production and agricultural incomes in Kenya, Tanzania, and Uganda.[16] Empowered women would not only receive agricultural extension services that increase their yields or income, but would also be able to influence what crops they can grow and the types of extension services they receive to ensure that the services meet their needs in both content and mode of delivery.

Measuring program success therefore requires that the indicators used to track progress are consistent with the program's reach, benefit, or empowerment goals and strategies. Participation counts indicate how well a program reached women, but not whether a program delivered benefits to or empowered women. To assess whether women benefited from a program, sex-disaggregated data on key outcomes of interest are necessary to compare the benefits that accrued to women versus men. However, even when women benefit, they are not necessarily empowered. To assess whether the program enhanced women's ability to make strategic life choices, programs must collect indicators on decisionmaking power and other aspects of empowerment, such as those captured in the Women's Empowerment in Agriculture Index (WEAI) or its project-level counterpart (pro-WEAI).[17]

To take another example, value chain interventions may reach women by including them in informational meetings, and often assume they will benefit if household incomes increase. However, the income from sales of produce often goes to men because women lack mobility and transportation to take produce to market. As a result, the extent to which women benefit from market development will depend on the degree of sharing and cooperation within the household. For example, a study in Ghana found that with the introduction of warehousing systems, produce that had been under women's control at the homestead was now under men's control, and women did not know how much grain their husbands had sold, or what happened to the income.[18] Empowering women in value chains would involve identifying women's constraints and the ways that women can gain control over income. First steps toward empowerment often include providing women with access to some form of financial account, whether through a bank, microfinance organization, or mobile money platform; ensuring that they can get produce to market, perhaps through local collection centers; and including women on contracts or warehousing receipt systems.

Rural electrification illustrates how programs can support both women and men in the process of rural transformation and contribute to women's empowerment as well as sustainable program outcomes. These programs would be more financially viable if women, as well as men, could use the energy for both productive and domestic uses. Electrification could also help reduce women's workload and provide lighting for girls' and boys' schooling. Ensuring that the appropriate investments are made requires that women's and girls' time is sufficiently valued—by women, their families, and communities.

REACH, BENEFIT, EMPOWER FOR POLICY

Adopting a *reach-benefit-empower* framework for projects is useful, but their impact will be sustainable only if women's empowerment is also supported at the policy level. For example, projects that create women's groups to save money or obtain access to microfinance loans will not have a sustainable impact if women cannot eventually "graduate" from microfinance to the formal banking system. Programs that give women microplots for agricultural production will not be sustainable if laws do not protect women's legal rights to land. Older women or women whose marriages have dissolved will not have secure rights to property if inheritance or divorce laws are inequitable.

Policies to reach, benefit, and empower rural women and girls can support rural revitalization because they support not only environmental sustainability and agricultural transformation, but also the development of more profitable nonfarm enterprises and better opportunities for women and men.

Relevant policies will go beyond those directly related to agriculture to address other constraints facing rural women. For example, policies that reduce the burden of women's family responsibilities—such as public support of childcare—will become more important as more women transition out of agriculture into the rural nonfarm sector. Social protection, such as health insurance, and better provision of primary healthcare could reduce the pressure on women as caregivers and help protect their assets. In

Bangladesh, for example, women's assets are often sold first to pay for unexpected health expenses.[19]

Policies that enable government services to *reach* women include those that improve women's access to infrastructure and information and that strengthen agricultural extension systems so that both men and women farmers are better served. This could include working through women's groups and hiring more male and female extension workers, as well as teaching extension workers how to better interact with male and female farmers.[20]

Policies that *benefit* women and girls provide access to public services that women and girls need (such as water supply and sanitation) or attempt to redress discrimination in schooling, access to the labor market, and access to financial services. A very successful example is the female secondary-school stipend program in Bangladesh, introduced in 1994 to improve rural girls' education. The program provided free secondary education and a cash stipend for eligible girls residing in rural areas. A study of the program's long-term impact found that it increased years of schooling for female students by 0.6 to 1.2 years, and that girls exposed to the program had lower desired and actual fertility.[21] This better-educated female workforce was more likely to be employed in the formal sector (like the country's garment industry) and less likely to work in the agriculture or informal sector. Stipend-eligible women also married more-educated husbands who had better occupations and who were closer to them in age. The children of eligible women were taller and heavier for their age. These findings suggest that government policies and programs to close gender gaps in schooling can have far-reaching effects later in life and into the next generation.

Policies and institutions that give women greater voice in governance and improve gender equality not only *empower* women and girls but also lay the foundation for sustainable rural revitalization. For example, the reform of the Ethiopia Family Code and its community-based land registration program set the stage for progress in gender equality in Ethiopia.[22] The revision of the country's marriage laws to give women a claim to family assets in the case of divorce or death helped increase investments in girls' schooling: mothers who perceived that they would receive an equal share of assets in case of divorce invested more in their daughters' schooling than those who expected a smaller share.[23] But policy reform alone is not sufficient to achieve women's empowerment—policies to support women must be complemented by activities that raise women's awareness of their rights and how to claim them as well as efforts to increase their participation in governance.

Such efforts, like quotas for women in leadership, show promise for improving the responsiveness of these institutions to women's needs and making sure that women's priorities are reflected in policymaking. For example, in India, after the passage of the Panchayati Raj Act requiring that 30 percent of representatives in local governments be women, villages with women leaders were found to have more publicly provided goods related to women's priorities, such as water supply and education. However, quotas for women in government bodies are insufficient to ensure gender-equitable service delivery without concomitant programs to build women's active participation and strengthen the capacity of service providers to work with women.[24] Quotas could be interpreted as ceilings for women, and having many visible women in the higher echelons of government does not necessarily mean that women have a voice at the grassroots level. It is also important to boost women's confidence to speak up and become more politically involved at the local level, such as through women's organizations like India's self-help groups.[25]

FROM COMMITMENT TO REALITY

Commitment to gender equality and women's empowerment is growing at the international and national level, but translating this commitment into reality is not straightforward. No single blueprint exists, because gender issues, and the constraints and opportunities that women and girls face, depend on the context. However, reforms that put greater resources under women's control, reduce their workloads, and provide greater space for women's effective voice in governance structures are important, and should be coupled with improved data systems and impact assessments to document changes in empowerment and the distribution of benefits. These reforms generally require going beyond top-down interventions. Formulating, implementing, and monitoring both programs and policies through the *reach-benefit-empower* lens can help assure that women's needs are met and that women have greater capacity to contribute to rural revitalization.

"Formulating, implementing, and monitoring both programs and policies through the *reach-benefit-empower* lens can help assure that women's needs are met and that women have greater capacity to contribute to rural revitalization."

CHAPTER 6

Environment
Revitalizing, Restoring, and Improving Rural Areas

CLAUDIA RINGLER AND RUTH MEINZEN-DICK

Claudia Ringler is deputy division director and **Ruth Meinzen-Dick** is a senior research fellow, Environment and Production Technology Division, International Food Policy Research Institute, Washington, DC, USA.

KEY FINDINGS

- Rural areas are the locus of essential ecosystem services for the planet, and rural dwellers are key custodians of these services; at the same time, rural livelihoods, including agriculture, contribute to loss of forests, groundwater depletion, land degradation, water and air pollution, and climate change.
- Challenges contributing to this environmental degradation include population and (unequal) economic growth and the resulting pressures on natural resources; associated poverty and inequality that leave people—particularly women—with few options; and lack of rights and poor pricing of resources, which hamper sustainable use.
- Restoring and improving the natural environment is a key component of rural revitalization, essential not only to improving rural living conditions but also to the health of the planet. Agriculture and its custodians—women and men farmers—are central to the revitalization and restoration of rural environments.

KEY RECOMMENDATIONS

- Invest in rural revitalization to create healthy and thriving rural areas and to provide safe food, clean water, climate change mitigation, and other key environmental services.
- Provide economic incentives to address environmental degradation. Possibilities include payments for environmental services, removal of subsidies that distort resource use, and well-defined, tradable rights for environmental "goods," such as clean water, and "bads," such as pollution.
- Invest in innovative practices and technologies, such as precision farming, small-scale irrigation, and communication technologies, that can increase agricultural yields and reduce environmental degradation.
- Support context-appropriate institutions to motivate and coordinate action among rural dwellers to address environmental issues. These institutions include land and other resource rights, local resource users' associations, and partnerships with national governments.

Large fish kills on Viet Nam's coast, acceleration of deforestation in the Brazilian Amazon, and growing risks predicted from global warming for biodiversity, natural resources, and agriculture are just some of the environmental problems affecting rural areas and, consequently, the whole world.[1] At the same time, the number of undernourished people increased to nearly 821 million in 2017, largely because of the often intertwined problems of civil conflict and climatic challenges that particularly affect rural populations.[2] To address these challenges, rural revitalization must go beyond economic progress as conventionally measured to also restore and improve the natural environment—not only for better rural living conditions, but also for the health of the planet. Agriculture has a key role to play in this revitalization of the environment.

Rural areas are the locus of key ecosystem services, including: provisioning services, such as the production of food and water; regulating services, such as control of climate; supporting services, such as oxygen production; and cultural services, such as spiritual and recreational spaces for both urban and rural dwellers (Figure 1). In fact, all of humanity depends on the health of the world's ecosystems in rural areas.[3] But rural environments and the services they provide are in danger: climate change, deforestation, soil degradation, water and air pollution, and solid wastes are growing threats to rural production, sustainability, and the well-being of rural and urban residents alike. Lack of rural infrastructure and services and poor economic opportunities compound these threats to the potential of rural areas. A comprehensive approach to creating sustainable rural systems, as recognized under the United Nations' Agenda 21, would ensure that ecosystem services are managed to meet present and future needs, while providing adequate services and economic benefits for everyone.[4]

KEY ENVIRONMENTAL TRENDS IN RURAL AREAS

Environmental trends in rural areas generally point to rapid and sometimes irreversible losses of key ecosystems and their functions, which in turn will affect social and economic outcomes for agriculture and rural areas.

FIGURE 1 Key functions and relationships affecting rural environments

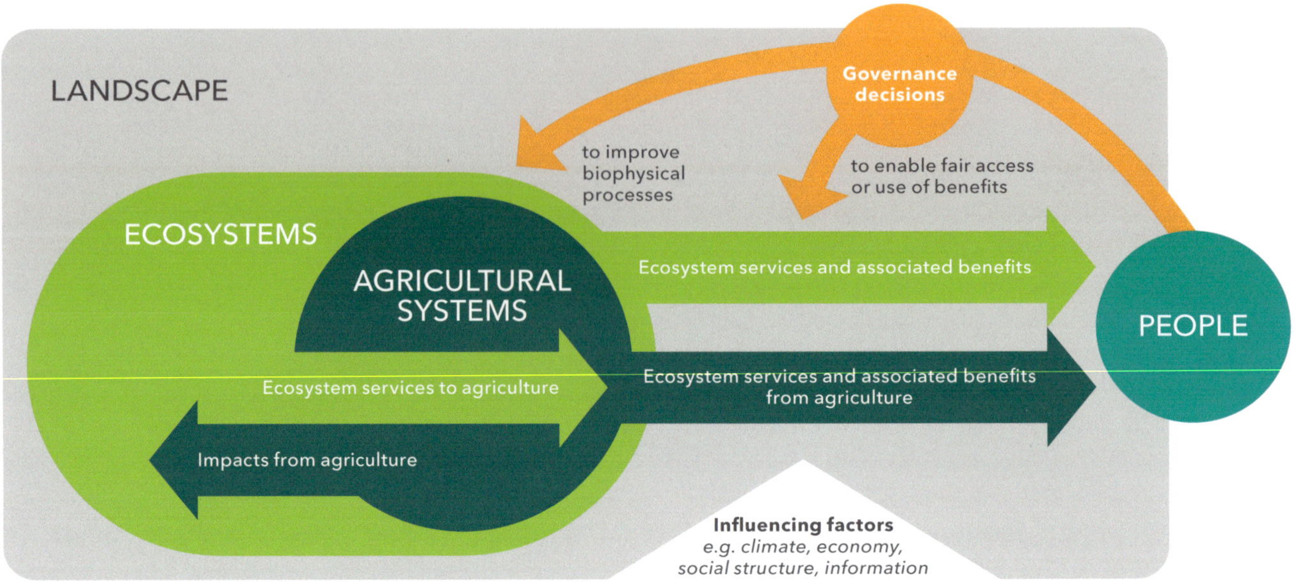

Source: Based on CGIAR Research Program on Water, Land and Ecosystems, *Ecosystem Services and Resilience Framework* (Colombo: 2014).

CLIMATE CHANGE. Global warming has already reached approximately 1°C above pre-industrial levels, with larger warming over land. Higher temperatures and extreme weather events disrupt agriculture and food systems, and have already contributed to increased hunger and malnutrition. As solutions for climate change mitigation in the energy sector are becoming available and relatively affordable, the largest challenges for both mitigation and adaptation strategies lie in the agriculture space, which contributes up to one-quarter of greenhouse gas emissions.

DEFORESTATION. Forests provide essential timber and nontimber products (such as food, fuel, housing, and paper), provide living space and health benefits, and harbor more than half the world's species. Tropical forests will be the key source of natural climate solutions for the next 10 years, chiefly through reforestation and avoided deforestation.[5] Despite solid evidence of the critical functions of forests, loss of tropical forests has been increasing for almost two decades, reaching 15.8 million hectares in 2017 (with total tree cover loss at more than 25 million hectares).[6] And while humans are to blame for almost all direct deforestation globally, their indirect contribution, through climate change–related disasters such as fires and storms, is also growing.

LAND DEGRADATION. One-third of the world's lands are considered moderately to highly degraded due to the erosion, compaction, waterlogging, salinization, acidification, and chemical pollution of soils, with erosion being the key challenge.[7] Many of these challenges are directly caused by agricultural production: tillage can compact the soil or expose it to erosion; irrigation with insufficient drainage leads to waterlogging and salinization; agrochemical applications contribute to chemical pollution; and ammonium-based fertilizers contribute to acidification. Addressing these challenges will be essential for increasing agricultural productivity and food security, particularly in Africa south of the Sahara.

WATER POLLUTION. Agricultural pollution is a growing threat to the world's water supplies, particularly in developing countries where it remains largely unaddressed. Excessive use of inorganic fertilizers and effluents from livestock production are rapidly increasing nitrogen and phosphorus runoff. While Brazil, China, India, and the United States currently have the largest agricultural pollution loads, going forward pollution growth is expected to be fastest in low-income countries.[8] Industrial water pollution also affects agriculture, fisheries, and rural drinking water. Without careful attention to water quality, rural economic transformations that increase processing and industrial production pose considerable risks to the environment and human health.[9]

GROUNDWATER MINING. Over the last four decades, extraction of groundwater has increased dramatically to support agricultural development in many parts of the world, including China, India, the United States, and parts of the Middle East. Today, more than a third of global irrigated cropland relies on groundwater.[10] Resulting groundwater depletion and degradation affect farmers, rural communities, and the environment through drying-up of springs and wetlands, salinization, and pollution. Availability of free or cheap electricity for pumping (most recently through solar power) contributes to the overuse of groundwater. Addressing this depletion is particularly challenging because groundwater cannot be easily observed and the many users of groundwater are often dispersed.

SOLID WASTE. With increased wealth, population growth, and urbanization, the generation of solid waste has grown exponentially in recent decades. Daily per capita generation of solid waste ranges from 0.1 to 4.5 kilograms globally, and while levels are highest in high-income countries, the rest of the world is catching up fast.[11] In low-income countries, where just over a quarter of waste in rural areas is collected (compared with almost half in urban areas), most waste is dumped in open landfills in rural areas and growth in non-managed waste disposal is faster than managed waste disposal. The impacts on public health and the environment have yet to be fully assessed.

AIR POLLUTION. Agriculture is a major contributor to air pollution: directly through emissions of ammonia gas from livestock manure and intensive cropland production, the burning of crop residues, and machinery operations; and indirectly through the burning of forests in shifting cultivation practices.[12] Ammonia causes soil acidification, eutrophication, and ground-level ozone and contributes to the formation of airborne particulates linked to respiratory disease and premature death. The rural dependence on burning wood, charcoal, crop residues, and dung—because of lack of access to clean energy sources—not only contributes to deforestation and soil depletion but also exposes approximately 2.8 billion people to household air pollution, with an estimated annual death rate of approximately 4.3 million people.[13] Rural and urban air pollution often combine in a deadly cocktail, as happens in wintertime in New Delhi, when pollutants from the burning of wood and rural crop residues are trapped in the city, adding to the pollution from vehicle emissions and local industries, and shaving around six years off the life expectancy of city residents.[14]

SOCIAL AND ECONOMIC DRIVERS OF ENVIRONMENTAL DEGRADATION

POPULATION GROWTH. By 2050, about 3.1 billion people are expected to reside in rural areas (compared with 6.7 billion in urban areas), but the share of rural populations will vary substantially by region and across countries. Only Africa south of the Sahara expects a substantial increase in total rural population, by 260 million people between 2020 and 2050, but that increase will be much less than that of the region's urban population, projected to grow by around 800 million people during the same period.

Population growth intensifies pressure on resources through higher demand for food and fiber, use of natural resources, and generation of waste. The demographic shift from rural to urban areas and the occupational shift from farm to nonfarm can have positive or negative implications for the environment. As agricultural labor becomes more scarce, labor-intensive approaches to resource conservation, such as terracing, and labor-intensive maintenance of irrigation become more difficult. The labor shortage is compounded in many areas where youth and men are likely to migrate or shift to nonagricultural enterprises, resulting in aging of the agricultural workforce and increased responsibilities for women. However, labor constraints can also create opportunities for environmental conservation, such as a shift from annual crop farming to more sustainable agroforestry and permaculture approaches.[15]

Agriculture remains the main employer in most rural areas, and as such is essential to rural livelihoods and rural economic growth in most developing countries. However, agricultural laborers face large safety and health risks, long working hours, energy-intensive work, and low compensation. Monitoring working conditions and remuneration in agriculture are important measures of economic and social sustainability that can also support the environmental functions of rural areas.

POVERTY AND INEQUALITY. As countries struggle to provide services for rapidly growing cities, rural areas are often left under- or unserved. Access to education, health, clean water, transportation, and electricity is dramatically lower in rural areas of developing countries compared with their urban counterparts. With low income levels and limited opportunities in more remote rural areas, it is not surprising that environmental degradation—as a result of reliance on fuelwood for cooking and heating or from using slash-and-burn agriculture, for example—is common. However, as incomes rise, environmental degradation tends to increase—as chemicals are overapplied due to increased availability without concomitant agricultural extension, or groundwater is overdrawn as a result of access to motor pumps.

In many countries, it is predominantly men who migrate from rural areas or shift to nonagricultural activities, while women take up an increasing share of agricultural work. This has implications for environmental conditions because women are often more labor-constrained and have less formal education, less control over assets, and less access to extension services. Women are also less likely to have recognized land rights, resulting in lower incentives and ability to invest in sustainable land management practices.[16]

RESOURCE RIGHTS AND RESOURCE PRICING. Property rights to land and other resources are important to provide resource users with the authority and motivation to make long-term investments in those resources. For example, land users—whether individuals or communities—without secure rights have little incentive to plant and care for trees, and in many cases might even be prohibited from doing so.[17] Lack of state recognition of land rights is especially problematic for collectively held resources such as pastures, forests, and riparian areas—all areas that are particularly important for environmental services—and for marginalized groups such as pastoralists, forest-dependent communities, and indigenous peoples. Even where household land rights are secure, women often lack recognized land rights.

IMPROVING RURAL ENVIRONMENTS

Rural areas are underfunded despite their essential contribution to human and planetary health.[18] Shifting demographics will likely further concentrate funding in urban areas and further displace funding from rural areas. But failure to invest in rural revitalization will likely harm both rural and urban areas, as urban areas rely on healthy and thriving rural areas for safe food, clean water, and energy, as well as the dilution of wastewater, disposal of solid waste, and important psychological and other health benefits yet to be adequately valued.[19] Ensuring the future of both rural and urban areas will require investments, policies, and institutions that directly support the environment and provide social services and economic opportunities that do not adversely affect the environment. These are discussed in other chapters of this report. Here we focus on complementary investments, policies, and institutions that benefit the environment, while also enhancing social and economic opportunities in rural areas. An example from China illustrates the need to consider rural revitalization from an environmental perspective (Box 1).

ECONOMIC INCENTIVES TO ADDRESS ENVIRONMENTAL DEGRADATION

Three types of economic incentives can help to address rural environmental degradation. Payments for ecosystem services are incentives for environmental management, generally in the form of financial compensation paid to landowners in exchange for implementing specific land management or other practices on their land (including removal of land from farming). Payments for ecosystem services are most commonly used for climate change mitigation, watershed protection, and biodiversity conservation (Box 2).

Subsidies for electricity, fuels, fertilizer, irrigation water, and other scarce or potentially environmentally damaging inputs lead to overuse, poor balancing of agricultural inputs, and often unintended secondary effects. For example, free access to electricity for pumps has led to widespread groundwater depletion in parts of India. As removal of such subsidies has short-term negative impacts on farm incomes, political opposition to their removal often arises. Replacing such subsidies with direct income support to farmers, or with other forms of support such as payments for environmental services, will be essential for rural revitalization.

BOX 1 The ups and downs of rural revitalization in China

China has made significant strides in reducing poverty and supporting agricultural intensification and rural development. This rapid improvement for rural areas has led to severe environmental degradation, however, which the country began to address in the late 1990s through large-scale investments in infrastructure, as well as through regulations and fines. Possibly the best-known environmental intervention the country has undertaken is the re-greening of parts of the Loess Plateau, an area subject to severe erosion. Between 1999 and 2008, the Chinese government invested around US$30 billion in the Grain to Green Program to retire and re-green about 9.3 million hectares of sloping land, with farmer participation. This led to substantial changes in land cover: woodlands increased by 4.9 percent, grasslands by 6.6 percent, and residential land cover by 8.5 percent, while farmland decreased by 10.8 percent. Soil conservation and carbon storage as well as grain production increased, although water yield (the amount of water "produced" by the watershed) declined.

Despite these positive developments, much remains to be done. For example, almost three-quarters of the river sections monitored in China were deemed unfit for human contact in 2002, and despite massive investments in pollution cleanup, 31 percent of monitored sections still remained unfit in 2015. High levels of air pollution in China have been shown to negatively impact mental health, perceived well-being, and cognitive development. And the impacts of injudicious use of antimicrobials on swine and poultry farms are yet to be fully understood. These human health costs of poor agricultural and natural resource management practices have not been factored into the economic growth figures of China or any other country.

See Y. Lü et al., "A Policy-Driven Large Scale Ecological Restoration: Quantifying Ecosystem Services Changes in the Loess Plateau of China," *PLoS ONE* 7, no. 2 (2012): e31782; M. Wang et al., "Rural Industries and Water Pollution in China," *Journal of Environmental Management* 86 (2008): 648–659; China Water Risk, "2016 State of Environment Report Review," June 14, 2017; X. Zhang, X. Zhang, and X. Chen, "Happiness in the Air: How Does a Dirty Sky Affect Mental Health and Subjective Well-Being?," *Journal of Environmental Economics and Management* 85 (2017): 81–94; X. Zhang, X. Chen, and X. Zhang, "The Impact of Exposure to Air Pollution on Cognitive Performance," *PNAS* 115, no. 37 (2018): 9193–9197; H. Yang et al., "Characterization of Multiple-Antimicrobial-Resistant Escherichia Coli Isolates from Diseased Chickens and Swine in China," *Journal of Clinical Microbiology* 42, no. 8 (2004): 3483–3489.

BOX 2 Payments for ecosystem services in Costa Rica

A well-known payment program (Pago por Servicios Ambientales) has been operating in Costa Rica since 1997 to address deforestation and biodiversity loss and to restore the hydrological services of the country's watersheds. The program is run at the national level and has been funded by various mechanisms–including, most recently, a fuel tax and a water tariff–that require users of key rural services, such as clean water, to pay to retain these services. This program led to an estimated increase in Costa Rica's forest cover of 0.4 million hectares by 2005, reversing earlier trends, and to a decline in carbon emissions. However, because the program provides a flat-rate payment to all participating landowners regardless of the ecological value of their land, substantial land of high ecological value was not enrolled.

See S. Pagiola, "Payments for Environmental Services in Costa Rica," *Ecological Economics* 65, no. 4 (2008): 712–724.

In addition to appropriate pricing of natural resources, well-defined, tradable rights for environmental "goods," such as water, or "bads," such as greenhouse gas emissions and water and air pollution, can motivate users to consume or pollute the resource more judiciously. A well-managed system of tradable rights makes users internalize the external costs of resource use or pollution, reducing incentives to degrade natural resources. The water market in Australia's Murray-Darling Basin illustrates how markets can generate significant economic benefits if scarce resources are carefully managed.[20]

INNOVATION AND TECHNOLOGIES

A number of farm- and landscape-based approaches can reduce environmental degradation, while also providing income opportunities that can reduce poverty and inequality. Many studies have shown that improved crop and water management, including use of a range of environmentally oriented technologies, such as enhanced water harvesting, conservation tillage, and small-scale precision farming, complemented by advances in breeding, such as heat and drought tolerance, can improve agricultural production while conserving scarce water and land resources.[21] Small-scale irrigation, for example, has been shown to double or triple crop yields in Africa south of the Sahara, reducing the need to expand into more fragile lands while also improving income, health, and nutrition outcomes, and—with the right supporting services—can benefit the rural poor.[22] Finally, access to information and communication technologies can support any and all technological innovations for environmental, social, and economic revitalization of rural areas.

REFORM OF INSTITUTIONS

While technologies and economic incentives are crucial for redressing environmental degradation, they must be supported by appropriate institutions to motivate and coordinate action. This includes legal recognition of the rights of rural men, women, and communities to land, water, forests, and fisheries. Land tenure security is shown, overall, to contribute to productive and environmentally beneficial investments in the land.[23] For example, land certification programs that register land in women's names in Rwanda and Ethiopia increased investment in soil conservation and tree planting by women-headed households.[24] Women's knowledge of their land rights has also been shown to increase investment in soil conservation and tree planting in Ethiopia and Uganda.[25]

Because most natural resource management has externalities with impacts beyond the individual farm or community, postive outcomes require coordination—among individuals, communities, and with government agencies. Local organizations such as water-user associations or forest-user groups can facilitate coordination by providing rules about who is responsible for contributing to the resource, who may use it, and in what ways. Monitoring and sanctioning mechanisms provide assurance that if one follows the rules, others will as well. But communities cannot undertake all resource management by themselves. Partnerships with government agencies are also required—for example, to regulate polluting industries. Such partnerships must support, rather than undermine, local involvement in resource governance.[26] For example, programs to increase community involvement in forest management in Tanzania have been shown to improve overall local governance, in that participants valued the greater control of their forest, regular access to forest products, and recognition from other villages and the government for their conservation efforts. At the same time, there is little evidence of improvements in incomes, owing to limited economic opportunities for forest use, and mixed evidence on the impact on forest cover compared with forests under government management.[27] Improvements in governance and local identity, along with incomes and environmental conditions, should be considered in any strategies for and visions of rural revitalization.

What is next for rural areas? We must begin with recognition of the role that improved natural resource management and joint support to the environmental, social, and economic functions of rural areas play in rural revitalization and their contribution to human and planetary health. We also must recognize the importance of agriculture in shaping environmental, social, and economic outcomes in rural areas and globally. As such, an assessment and valuation of the key ecosystem functions of rural areas—whose improvement will create social and economic benefits—must be the first step in any effort for rural revitalization.

Once these values are clear, entry points for improved rural environmental outcomes can be found among economic incentives, innovations, and institutional change, but these need to be tailored to the specific socioeconomic and environmental contexts of particular rural areas. To engage rural residents as custodians of valuable resources, their rights to resources should be recognized in law and practice. As women take on greater roles in agriculture, it is essential that they receive the information and support needed to adopt sustainable, climate-resilient agricultural practices. While the future of rural areas is under threat, there are many pathways toward more sustainable development. It is upon us—particularly in our activities and investments related to the agriculture sector—to ensure that these pathways lead to improved environmental outcomes and better lives for everyone.

"Rural revitalization must go beyond economic progress as conventionally measured to also restore and improve the natural environment–not only for better rural living conditions, but also for the health of the planet."

CHAPTER 7

Renewable Energy
Bringing Electricity to Revitalize Africa's Rural Areas

CHANNING ARNDT

Channing Arndt is division director, Environment and Production Technology Division, International Food Policy Research Institute, Washington, DC, USA.

KEY FINDINGS

- Energy is crucial to achieving the Sustainable Development Goals and ensuring durable rural growth and development. But almost one billion people, most living in rural areas of Africa and South Asia, still lack access to electricity.
- Modern electrical systems, particularly solar power, are making it easier than ever to meet the energy needs of dispersed rural populations in developing countries.
- Costs of solar systems have fallen dramatically and, together with the high solar potential in many developing countries, these systems offer a host of rural livelihood and development possibilities. As a result, genuine potential exists to ensure access to electricity for all by 2030.
- Electricity brings numerous benefits—from lighting to sanitation to access to information—and even the potential to increase consumption of healthy, nutrient-dense foods where irrigation and refrigeration are introduced.
- Expanding solar power use in Africa and other developing regions will require new approaches to energy planning and implementation to reach dispersed rural populations.

KEY RECOMMENDATIONS

- Provide public support to increase electricity access in remote rural areas through policies to promote investment and competition among providers of distributed solar or other renewable systems.
- Deliver packages of support—access to electricity plus the means to use it, such as lighting and refrigeration—to generate larger development benefits for rural communities.
- Account for the impacts of adoption of new technologies within households, particularly on women, including the impact on workloads.
- Be aware of potential unintended consequences of widespread technology adoption, such as more rapid depletion of groundwater resources associated with low-cost electrical pumping.
- Consider the potential for enhancing rural development and serving urban electricity needs when selecting sites for power generation.

It is difficult to envision hundreds of millions of absolutely poor people escaping from poverty without access to energy, notably electricity. Energy access is now broadly recognized as crucial for achieving nearly all the Sustainable Development Goals (SDGs).[1] In recognition of the critical roles that lighting, irrigation, refrigeration, and other energy uses must play in durable processes of rural growth and development, energy access is highlighted with its own SDG (7.1), which aspires to "universal access to affordable, reliable, and modern energy services" by 2030.

Unfortunately, fossil-fuel-based energy systems serve rural areas in developing countries, particularly in Africa, with difficulty. Large, centralized power systems with complex operating requirements do not easily reach widely dispersed and poor rural populations. For developing countries, the enormous investment required to deliver electricity to these rural populations has hindered progress in provision of energy services. As a result, more than one billion people, or about one person in seven, lacked access to electricity in 2014. Nearly nine out of ten of those resided in rural areas, primarily in Africa and South Asia.[2] Africa south of the Sahara, where populations are both widely dispersed and particularly poor, accounted for about 60 percent of those lacking access to electricity.

But modern energy systems are changing in ways that make meeting the energy needs of dispersed rural populations in Africa and other developing regions easier. Steep declines in the cost of solar power and other distributed (that is, noncentralized) energy systems are creating new possibilities for electrification. This is especially true in sunny places, such as broad swathes of rural Africa and South Asia. With this new potential of renewables, a major obstacle to rural electrification—the cost of delivering reliable energy to rural inhabitants—is receding.

This is good news. Reasonably reliable, low-cost renewable energy systems are creating a host of livelihood and development possibilities for remote and rural areas. As yet, however, the policies and institutions needed to take advantage of these new possibilities, as well as to address their potential unintended consequences, are in their infancy. In this changing energy landscape, fresh thinking about rural development priorities, policies, and investments for powering homes, farms, and industries promises transformative change for rural areas. This chapter looks at the broad implications of a changing energy landscape for electricity access in developing countries and, in particular, at the role that solar power can play in rural Africa.

A CHANGING ENERGY LANDSCAPE

Falling costs and prices of solar power and other renewables are defining the new energy landscape. Costs (measured as levelized costs[3] excluding all subsidies) of solar power generation using photovoltaic (PV) systems have dropped dramatically—by a factor of three or more since 2010 (Figure 1). Costs have also fallen relative to other generation sources. For example, by 2017, the costs of community solar PV systems were in the same range as utility-scale coal-fired power (Figure 2).

Figures 1 and 2 provide cost estimates for the United States excluding subsidies or other policy impacts. Actual prices paid, which include the effects of policies, exhibit similar trends. At the global level, the average price paid at auction by utilities to purchase onshore wind power and, especially, solar PV at grid scale has declined dramatically (Figure 3). Power providers in 2017 were bidding to deliver power in 2018, 2019, and 2020, so these data effectively project prices into the future. Notably, the average price of solar PV falls below that of onshore wind in 2019 and continues downward.

Overall, the three figures point to the following qualitative conclusions:

- Generation costs for renewables (using a levelized cost of energy metric) have been falling rapidly.
- Declines in the cost of solar PV generation have been particularly rapid.
- Unsubsidized cost levels are in many circumstances comparable to and sometimes below those of traditional fossil fuels.
- Cost declines seem likely to continue.

Two legitimate concerns should be mentioned with respect to these qualitative conclusions. First, levelized costs for variable renewable energy (as shown in Figures 1, 2, and 3) can be misleading from an energy systems perspective. Power generation from solar

FIGURE 1 Real cost of solar photovoltaic systems in the United States

Source: Based on National Renewable Energy Laboratory, *U.S. Solar Photovoltaic System Cost Benchmark,* Technical Report NREL/TP-6A20-68925 (Golden, CO: 2017).

Note: Costs are for utility-scale photovolatic fixed tilt 100 MW systems.

and wind systems is variable—dependent on the sun and wind—and generation cannot be adjusted to meet increases in demand. In the jargon, wind and solar systems are not "dispatchable." Matching power supply with power demand is a major challenge to incorporating large volumes of variable renewable energy into existing power grids. When variable renewables constitute only a small share of total power generation and the other sources are dispatchable, then variation in renewable power output is easily accommodated. However, with a large share of variable renewables in total output, matching power supply and demand becomes more complex.

For poor rural zones of Africa and South Asia seeking to establish mini-grids using solar power, this concern may not loom as large. For areas that have had no power (or highly unreliable power), even a system that provides power only when it is sunny is an improvement, and variability is a less pressing problem. When solar power is not available, such as at night, options include batteries and generation using bioenergy, diesel, or even wind power. For

FIGURE 2 Comparison of levelized energy costs for the United States, 2017

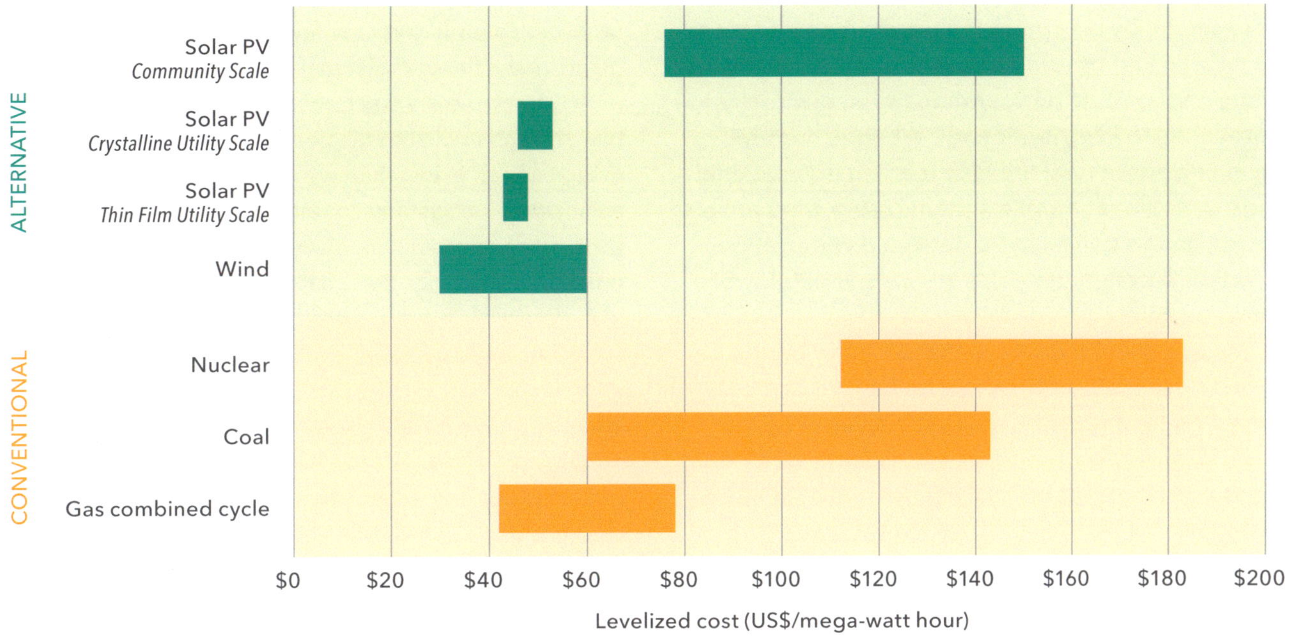

Source: Based on *Lazard's Levelized Cost of Energy Analysis*, Version 11.0, November 2017.

FIGURE 3 Cost of renewable energy at auction, global

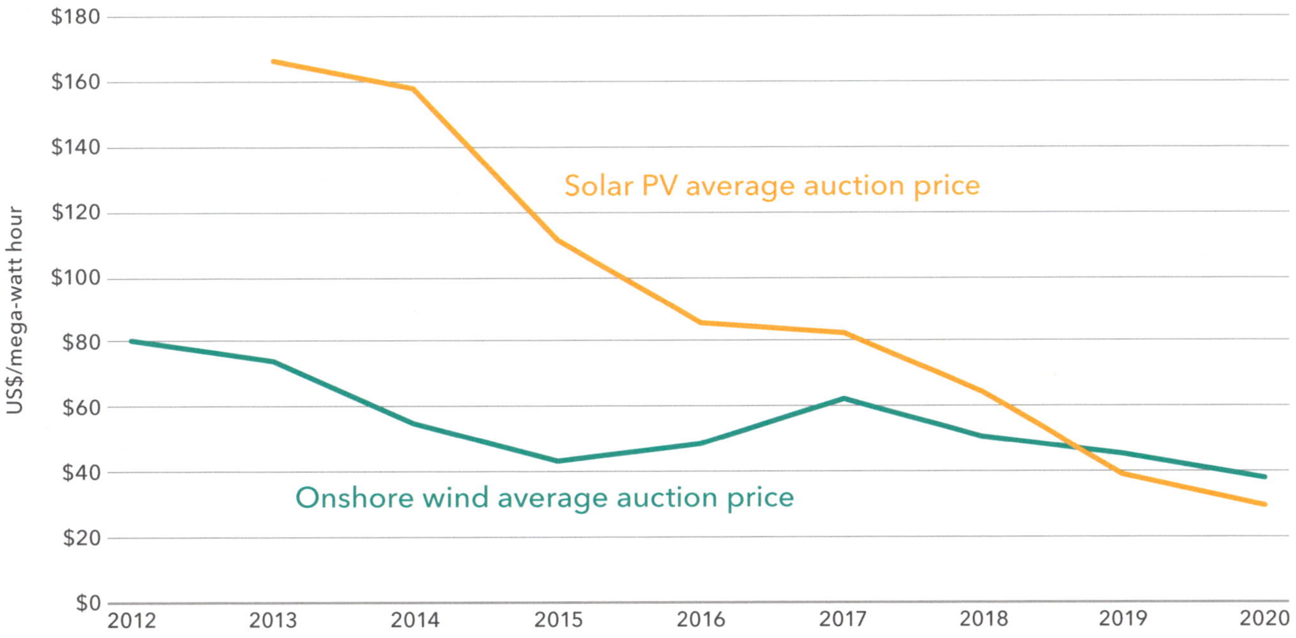

Source: International Energy Agency, *Renewables 2017: Analysis and Forecasts 2022*, Market Report Series (IEA/OECD: 2017). Reproduced with permission.

some important electricity uses in rural zones, such as water pumping for irrigation and refrigeration, variations in power output can be accommodated, for example, by only pumping water when electricity is available and by using a combination of ice, insulation, and variable refrigeration (when electricity is available) to keep items cool. Whether supplying mini-grids in rural areas or feeding into national grids, variable renewable systems using wind or solar energy generation must be designed and used in ways that accommodate this inherent variability of power output.

Second, the figures refer to the United States (Figures 1 and 2) and to the world (Figure 3). Costs in rural Africa and South Asia can be expected to differ. However, if experience with cellphone technology is a guide, costs are likely to be similar in these developing regions. In fact, for many locations in developing countries, solar radiation endowments are far superior to those of developed countries, which would help to lower costs. For example, a comparison of solar resources in South Africa and Germany, which gets about 7 percent of its electricity supply from solar power sources, shows that South Africa's solar potential is uniformly relatively high and Germany's relatively low.[4] In fact, the *worst* sites in South Africa yield more solar potential on an annual basis than the *best* sites in Germany. While the details will vary from country to country, the broad picture of relatively abundant solar resources across rural zones in Africa and South Asia holds.

Variability in power output and variations in potential across locations notwithstanding, the implications of the rapid and continuing drop in costs of renewables are already substantial. On a global basis, investment in renewable energy generation systems in 2017 reached US$280 billion, far exceeding investment in fossil-fuel-based power systems. In the United Kingdom, low-carbon-emissions sources of energy generation now account for about 50 percent of generation, up from about 20 percent in 2007, driven by a quadrupling of renewable generation over this period. Coal-fired generation in the United Kingdom fell from almost 40 percent of generation to less than 10 percent over the same period.[5] In South Africa, coal-fired generation predominates, but both solar and wind have strong potential. Three separate analyses have estimated South Africa's least-cost energy mix as containing about two-thirds variable renewables. In the studies, this least-cost mix would roll in over time, with renewables replacing coal-fired generation as coal plants reach the end of their useful lives.[6]

If renewable energy generation is now competitive enough in cost to displace fossil-fuel generation in centralized grids, then we can expect that local solar-powered systems in remote rural areas should be even more competitive, given that these systems have much lower distribution costs. The continuing declines in costs are also noteworthy.[7] As people in developing countries learn to install and use solar systems at various scales, further cost declines are likely to follow. Recent studies have found that existing solar power systems are economically viable in developing countries at prevailing costs.[8] As prices fall further, solar power is likely to become economically viable across a greater range of circumstances.

ENERGY AND RURAL NUTRITION

Energy, especially electricity, is critical for development, including for lighting, irrigation, and refrigeration. To this list, one could add numerous other uses, such as clean drinking water, access to information, and sanitation. Chapter 1 of the World Bank's *State of Electricity Access Report* provides a detailed list of benefits of electricity access.[9] Rather than reiterate these benefits, focus is given here to the implications of energy access for production of nutrient-dense foods and, consequently, potential implications for nutrition.

A recent comparative analysis of the cost of calories from different food sources across countries shows the relatively high cost of nutrient-dense foods in poor regions.[10] Relative to the least-cost staple food, the cost of calories from nutrient-rich foods—such as cow's milk, chicken eggs, and meat—is typically three to four times greater in poor countries than in high-income countries. For example, in high-income countries, a calorie from a chicken egg costs roughly three times more than a calorie from rice or wheat bread. However, in many parts of Africa, a calorie from a chicken egg costs more than nine times as much as a calorie from the local staple food crop. Similar conclusions hold for fruits and vegetables.

In many parts of Africa and Asia, households suffer inadequate dietary diversity, at least partially reflecting the relatively high cost of nutrient-dense foods. The micronutrient deficiencies resulting from inadequate dietary diversity are associated with elevated rates of stunting (low height-for-age) among children and potentially permanent negative impacts on cognition. This suggests that improving diets could yield substantial development returns and that reducing prices of nutrient-dense foods should be a policy priority.[11]

What is driving these high prices? Nutrient-dense foods such as fresh milk, eggs, meats, fruits, and vegetables are typically perishable, requiring refrigeration and proper handling. As a result, in poor country settings with limited cold storage and inefficient transportation, these foods are rarely imported or traded over long distances. The clearest route to reducing prices is to increase the volume and efficiency of local production. Here, energy also plays a key role. Production for local consumption is likely to require electrification both to support irrigation, often needed for efficient production of fruits and vegetables, and for refrigeration of perishables.

These electrification-production-nutrition linkages add to the already compelling case for striving toward universal access to electricity. The distributed nature of renewable energy sources, especially solar, also buttresses the logic for a focus on relatively small-scale distributed irrigation systems in developing regions.[12]

ACHIEVING NEAR UNIVERSAL ACCESS

Despite the clear potential of solar power systems, these systems will not expand rapidly in rural areas unless other obstacles to universal access are addressed. Rural dwellers in developing countries who lack access to electricity are typically poor and have limited access to credit. As a result, their ability to pay (effective demand, in economic terms) for modern energy is low and could remain low for a long time. Even were electricity provided for free, demand would likely be limited because these rural residents possess few (if any) appliances or other assets that use electricity. Nor do they have funds to purchase these assets.

What can be done to promote universal access? Future gains in electricity access are contingent on adequate public support in developing countries. For solar power to make a substantial contribution, policies will need to take account of the nature of these energy systems. Important differences between distributed renewable systems and centralized fossil-fuel-based systems require a new approach to public support and a new institutional framework for energy generation. Four differences are highlighted here.

First, the economies of scale inherent in traditional fossil-fuel-based systems mean there are few players in generation and distribution, creating a monopoly-like situation that argues for substantial state involvement, either through direct industry ownership or regulation, to limit market power and lower prices. In contrast, distributed solar power systems can be constructed and operated by many players. State involvement is needed not to regulate a lone or limited number of providers but to foment competition across an array of providers who can help extend electricity to poorly served areas. At a minimum, the model of a monopoly energy provider delivering services to rural communities should be rethought.

Second, with expanded possibilities for delivering electricity to dispersed rural communities should come expanded effort to deliver packages of support that include both electricity access and the means to use it. Rural electrification is only beneficial if it can be used to provide lighting, heating, cooling, pumping, and other productive services—it is a means to achieving development goals. A focus on packages of support opens the possibility of attaining much larger development benefits much more quickly. Along with the expansion of electricity access and use comes the need to be vigilant for unintended consequences. For example, solar power and electrical pumps can drive down the cost of pumping water dramatically, and without appropriate institutions in place to govern resource use, groundwater may be rapidly depleted. Also, disposal of spent solar cells and batteries raises environmental concerns.

Third, the selection of locations for renewable power installations will differ from traditional systems. Renewable systems not only offer new possibilities for local rural electrification, they can also supply power to the main electrical grids. Historically, location of fossil-fuel-based generating assets has been driven largely by fuel acquisition costs. For example, coal-fired power plants are frequently located near coal mines, eliminating the need to transport low-quality coal. Likewise, the location of grid-scale solar or wind systems will be driven at least in part by the location and properties of renewable energy endowments (reliability and quantity of sun and wind). Considerable incentives also exist to distribute utility-scale renewable power generation installations across space. Spatial distribution of utility-scale renewable assets creates a portfolio effect—with multiple locations, it is likely that the sun will be shining or the wind blowing somewhere. This portfolio effect helps to stabilize total power output from variable renewable sources and thus eases the problem of matching supply with demand. The costs of transmission to sources of demand must also be factored into site selection.

The potential for leveraging rural electrification for rural growth should enter into this calculus. In West Africa, for example, populations are concentrated along the coastal south, but solar energy potential is greater inland toward the Sahelian zones in the north. A legitimate question might be: Is it better to generate power near coastal cities for low distribution costs, or to build utility-scale solar power in the north and invest in transmission to deliver it to the south? Answering this question requires detailed location-specific analysis. If the utility-scale solar and associated transmission offer a low-cost means to contribute to rural development in the north, this benefit should be taken into account. And, if rural electricity demand can be modulated cheaply and easily, then the addition of rural zones not only can support rural development but also can help match electricity supply with demand.

Finally, technology adoption and the implications of technology once adopted are mediated by social circumstances. For example, a study in Tanzania and Ethiopia shows that access to small-scale irrigation resources does not guarantee success in improving dietary diversity.[13] Looking at a sample of households in both countries, the study found that irrigation was associated with improved dietary diversity in Ethiopia; however, the association between irrigation and dietary diversity was not statistically significant in Tanzania. These kinds of results highlight the need for understanding the impact of technology adoption within households. Other recent research on small-scale irrigation finds that who controls a technology within a household helps determine not only whether the technology will be adopted but also how the benefits and costs of adoption are distributed within the household.[14] Some evidence shows that solar-powered pumps could be a promising technology for women, especially where the pumps reduce both domestic and field labor requirements.[15] Pumps located near the household, where women exercise more control, are also preferred by women, according to a study in Ethiopia.[16]

Overall, the good fit between the technical characteristics of solar power systems and the dispersed populations and sunny environments that characterize many rural areas of developing countries argues for adopting the new approaches and undertaking the necessary institutional transformations. There is also reason for optimism. For 25 years, the number of people living without access to electricity in Africa south of the Sahara increased steadily—from 400 million in 1990 to 600 million in 2015. A decline finally registered in 2016.[17] Recently, the net number of people gaining access to electricity on a global basis has been about 120 million people per year, with solar power systems playing a substantial role.[18] Some simple back-of-the-envelope calculations suggest that near-universal access to electricity by 2030 is possible. Assuming that the population lacking access in 2017 was one billion people, the rate of population growth in this group is 3 percent per year, and the net gain in access to electricity is 100 million people per year (less than the net gain of 120 million people per year registered recently), then universal access could be attained by 2030, despite the headwinds created by rapid population growth. Maintaining net gains in electricity access of 100 million people per year will not be easy, but we have better tools than ever before for achieving this goal.

POWERING CHANGE

Many developing countries have promising endowments in renewable resources, are facing rapidly growing demand for electricity, and are unencumbered by legacy fossil-fuel generation systems. As the costs of wind and solar technologies drop below nonrenewable competitors, developing countries are positioned for rapid uptake of renewable energy systems. The case of South Africa, which has a relatively well-established electricity grid, provides an example of the transformation made possible by coupling strong renewable energy endowments with detailed energy analysis. This combination has led to a sea change in perspectives on least-cost power systems, and variable renewable energy is projected to replace coal as the main power source in South Africa. In countries where rural zones are poorly served by the traditional grid, distributed renewable energy systems—mini-grids, micro-grids, and grid extension—are making a rapid transformation in energy access and use possible.

In reviewing the promises and challenges of the renewable energy revolution for revitalizing rural areas, five points stand out.

- **TECHNICAL CHANGE.** A paradigm shift in rural electrification is occurring. Distributed solar power systems, often supplemented with batteries or other forms of generation, are transforming electrification and development possibilities.
- **ENERGY NEEDS.** The key role that energy, notably electricity, will play in achieving the SDGs and making rural areas more productive and livable is widely recognized.
- **NUTRITION LINKAGES.** Stunting and other manifestations of poor nutrition reflect in no small part high relative prices of nutrient-dense foods in low-income environments. Producing nutrient-dense foods locally could potentially reduce prices and increase availability, but this will require access to electricity for irrigation and cold storage.
- **ENVIRONMENT.** The availability of relatively inexpensive and reliable power opens new possibilities, not all of which are environmentally benign. For example, increased pumping of groundwater risks rapid depletion of aquifers. In addition, renewable systems bring with them their own innate environmental challenges.
- **PUBLIC SUPPORT AND INSTITUTIONAL FRAMEWORKS.** While the potential for powering rural transformation has never been greater, realizing this potential requires both continued public support and a rethinking of how this public support is delivered, including development of appropriate institutional structures and linkages to other aspects of rural revitalization.

Today's renewable energy revolution is providing valuable tools for enhancing growth and development prospects in developing countries while meeting environmental objectives at global and local scales. Accelerated research efforts and fresh thinking will help us to realize the potential of the ongoing energy revolution.

CHAPTER 8

Governance
Making Institutions Work for Rural Revitalization

KATRINA KOSEC AND DANIELLE RESNICK

Katrina Kosec and **Danielle Resnick** are senior research fellows in the Development Strategy and Governance Division, International Food Policy Research Institute, Washington, DC, USA.

KEY FINDINGS

- Three aspects of governance are critical for rural revitalization:
  - Appropriate and predictable laws and regulations—especially a legal framework to protect property rights and to create incentives to support business—are fundamental to economic growth and development.
  - Effective policy implementation and enforcement—which depend on the government having adequate technical skills, financial resources, and appropriate organizational structures—are essential to realize policy goals for rural areas.
  - Accountability—from frontline civil servants to elected politicians—is key to ensuring that governments respond to the needs of the poor. Competitive elections, access to information, and transparency help citizens hold their governments accountable.
- Devolution of governance to a subnational or local level can improve government responsiveness to local needs when responsibilities are matched with adequate funding and mechanisms to ensure government accountability.
- The information revolution offers new tools for improving governance for rural revitalization, from informing citizens to facilitating citizen policy inputs to monitoring implementation.

KEY RECOMMENDATIONS

- Establish an enabling and predictable regulatory environment to stimulate private sector investment and engagement for rural revitalization. Identifying enabling environments and types of business-government relations that catalyze investment is a first step.
- Build state capabilities and establish incentives for better service delivery, such as pay-for-performance and delivery units, to improve policy implementation at the national and subnational levels.
- Facilitate the information revolution to promote engagement of citizens with one another and with politicians and governments. Using information and communication technologies to support competitive elections, provide accurate information to citizens, and increase transparency is essential for rural prosperity.

Pursuing the many policy goals necessary for rural revitalization in developing countries—such as generating employment, connecting small towns and peri-urban areas with larger cities, attracting private investment, and providing public services—requires a deep understanding of the underlying governance landscape. Too often, ambitious, pro-poor policy reforms are stymied by entrenched political interests, poor intersectoral cooperation, and inadequate technical skills on the part of government bureaucrats. The centrality of governance for agricultural growth and food security was reaffirmed recently by a variety of major international donors and regional actors.[1] However, as interest grows in enhancing rural-urban linkages to both revitalize rural areas and add value to the food system and development process, the array of necessary policy interventions becomes more complex. Governance challenges too can become more pronounced, requiring coordination across multiple subnational actors and numerous ministries, including ministries of agriculture, urban development, environment, education, health, labor, and commerce.

How and why is good governance critical for efforts to revitalize rural areas? Given that the concept of governance has "many proprietors and many varieties of definition and explanation," we focus on just three relevant components of this broader concept.[2] First, we examine the regulatory dimensions of governance—that is, the array of laws and rules that shape the enabling environment for attracting investments across the rural-urban spectrum and contribute to policy coherence. While rigid regulatory regimes can sometimes stifle innovation and deter productivity, they may also safeguard the rights of marginalized groups and provide needed environmental protections. Alternatively, weak or absent regulation can generate uncertainty for citizens and investors alike.

Second, since ambitious laws and regulations relevant to revitalizing rural areas are meaningless in the absence of implementation and enforcement, we also focus on state capabilities. These include the technical skills and financial resources of bureaucrats at all levels as well as the overall institutional structure of the bureaucracy—all of which ultimately determine how well governments prioritize and deliver services.

Finally, we turn to the interface between states and citizens, which is captured by the range and effectiveness of accountability mechanisms. Such mechanisms, which are increasingly diverse due to advances in information and communication technologies (ICT), not only allow for the monitoring and evaluation of government policies and investments, but can also help ensure that policies and expenditures genuinely respond to the needs of the poor. In discussing each of these three governance dimensions, we review how they affect rural revitalization and highlight novel approaches that some countries are using to strengthen these areas.

LEGAL AND REGULATORY REGIMES

Legal and regulatory regimes shape the decisionmaking environment for individuals, businesses, and the public sector. While many development outcomes are affected by factors beyond government control, laws and regulations are exclusively within the mandate of governments and therefore offer one of the most accessible entry points for generating reforms necessary for rural revitalization.

Property rights, for both physical assets and intellectual property, are well recognized as fundamental to growth and development. Effectively upholding these rights can also reduce conflict and polarization. Insecurity of land rights in particular, which can result in land disputes, illegal land evictions, and suboptimal investment decisions, impedes agricultural productivity, food security, and economic growth.[3] In contrast, increasing tenure security boosts investments in land and farming technologies, thus contributing to greater food security.[4] The importance of land tenure and rental laws, even for welfare and migration, is highlighted by natural experiments and policy analysis in a diverse set of countries. For example, in Kyrgyzstan, children in regions that initiated land privatization soon after the collapse of socialism grew taller than did children in later-privatizing regions, owing in part to higher household food production and consumption.[5] In Ethiopia, migrants from rural areas are predominantly "pushed" from their homes rather than attracted by an urban "pull." Youth who receive small or no land inheritances often migrate away from rural areas, but this "push" factor is substantially muted where land rental markets are vibrant.[6] Similar findings have emerged in India, where rental restrictions reduce productivity and equity.[7] Together, these examples suggest that amending tenure and rental laws is one way to retain and attract youth to rural areas.

An enabling regulatory environment is also needed to promote agricultural productivity and encourage agro-industry, which is viewed as a critical means of generating rural jobs and stimulating economic transformation, especially in Africa.[8] The creation of the Enabling the Business of Agriculture index by the World Bank has been an important step in providing a measurement of countries' regulatory environments. This index assesses a country's regulatory setting in eight domains necessary for agribusiness to thrive (seed, fertilizer, machinery, finance, markets, transport, water, and ICT) and allows for comparisons over time and across countries.[9] By providing a transparent, comparable metric by which governments can gauge their performance, the index aims to spur lagging governments to pursue reforms needed to attract investment in agriculture and agribusiness. Foreign exchange, trade, and phytosanitary regulations are equally important, as they shape the incentives of private actors to invest in certain sectors and connect rural citizens to domestic and international markets.

In some cases, the private sector may be better equipped than the public sector to provide social and agricultural services—either alone or through public-private partnerships. A study covering 39 African countries over 25 years found that allowing private sector participation in the piped water sector significantly improved child health by lowering the rate of diarrheal diseases.[10] Research from Argentina shows that the health benefits of water privatization are largest in the poorest areas, suggesting that the most disadvantaged people may enjoy the greatest benefits.[11] In Indonesia, outsourcing the delivery of targeted food transfers to the private sector reduced operating costs without sacrificing quality and, when there was sufficient competition among private firms, even lowered the prices citizens paid.[12] The benefits of collaboration with the private sector cannot be realized, however, without establishing legal provisions ex ante that allow for, and define the scope of, private sector involvement in national economies. For instance, laws regarding taxation, labor, and land use can make investment attractive for private sector firms, encourage competition among firms to provide services, and protect citizens' rights where needed. And effective regulation of private firms can ensure that firms do not abuse workers or the environment and that citizens profit from their engagement.

While recognizing the importance of the legal and regulatory framework for creating an enabling environment for rural revitalization, two caveats are warranted. First, regulatory frameworks must be predictable; even more than the substance of regulation, frequent changes can undermine the ability of investors to plan and scale up operations—which tends to drive away investment. Second, frameworks should be revisited periodically to ensure compliance with best practices and compatibility with contemporary circumstances. For instance, as intellectual patents for pesticides have expired, the Sahel region has faced an influx of more than 100 varieties of herbicide. Weak and outdated regulations mean that many of these are counterfeit, resulting in reduced sales for legitimate private companies and increased presence of toxic chemicals in the food system.[13] The nonfarm, informal food trade sector is another place where updated regulation is needed. Many countries still apply colonial-era anti-vending laws that penalize informal trade, thereby undermining the ability of microenterprises to expand, limiting social mobility, and reducing access to affordable food for the poor.[14]

REINVIGORATING STATE CAPABILITIES

Despite the fundamental need for clear laws and regulations, regulatory frameworks are only effective when properly implemented and enforced. Unfortunately, many well-intentioned policies are either not implemented as intended or at all. Along with insufficient political will, weak state capacity is frequently blamed for poor policy implementation in developing countries.[15] Addressing such weaknesses requires going beyond traditional training courses for civil servants to instead identify faults in incentive structures for policy implementation that can result from problematic organizational structures, inadequate knowledge on the part of civil servants and decisionmakers alike, or limited financial resources.

Public sector organizational structures may be weak at multiple junctures. For example, "street-level" bureaucrats at the frontline of service delivery may exercise their own discretion in policy implementation, which may lead to outcomes on the ground that deviate from the original intentions of the policy.[16] In some cases, this discretion facilitates corruption, and since bureaucrats are not elected, citizens cannot hold them accountable for improper policy implementation.[17] Poorly adapted organizational structures are another source of weakness; developing country bureaucracies often practice "isomorphic mimicry," meaning they follow the practices and adopt the administrative structures of developed countries, even if these are contextually inappropriate.[18]

INCENTIVIZING CIVIL SERVANTS

One means of overcoming deficiencies in organizational structures that limit rural revitalization is to incentivize civil servants, at all levels, to deliver relevant services. Pay-for-performance contracts and performance-based financing are common approaches to enhance delivery of social services and reduce corruption among frontline service providers, such as teachers and health workers. In the health sector, this approach has been introduced in more than 30 developing countries.[19] A number of studies suggest that pay-for-performance can, under the right conditions, help to reduce corruption and absenteeism.[20] Other encouraging outcomes have also been documented. For example, in rural Uganda, a new incentive system that rewarded teachers for the individual improvements of each of their students increased student attendance rates and improved test scores for students in schools with books.[21] Similarly, a recruitment experiment in Mexico found that civil service candidates were more likely to choose positions in poorer communities if they received higher wages.[22]

Performance contracts, which commit bureaucrats to either delivering services or risking sanctions from their superiors, are a similarly compelling means of reorienting incentives, with implications for overall job security and public censure. One example is Rwanda's performance contract approach, known as *imihigo*. In precolonial days, the tradition of *imihigo* involved leaders vowing publicly to accomplish certain deeds or be publicly shamed if they failed to deliver. The modern version aims to ensure that districts deliver services in alignment with national goals outlined in Rwanda's Vision 2020 and its economic development and poverty reduction strategy. Central agencies and district governments negotiate measurable targets around national priorities, such as building schools and marketplaces, paving roads, and electrifying villages. The districts are audited annually by a local policy research institute and assigned scores to indicate the share of activities completed.[23] The achievements of the *imihigo* system, by district and sector, are then made publicly available every year on the website of the prime minister's office. This system of performance contracts has been credited with guiding the impressive infrastructure investments throughout the country since the end of the genocide.[24]

"Delivery units" integrate the concept of performance contracts but aim to improve intersectoral coordination, cultivate a service mentality, enhance timely information flows, and meet priority targets within a specified time. In concrete terms, delivery units consist of small teams of experts whose offices are usually located within the office of the chief executive (president, prime minister, or governor) or the finance ministry, and who coordinate activities across sectoral ministries. This ensures high-level political commitment to the delivery unit's work and provides it with access to decisionmakers at the ministerial level to help identify bottlenecks and facilitate information flows.[25] Typically, delivery units aim to translate agreed-upon sectoral goals into measurable targets and involve performance contracts between a president or prime minister and senior cabinet officials, who in turn have an incentive to ensure that their subordinates deliver. By creating metrics for successful service delivery that go beyond budgeted expenditures, and receiving frequent updates on these priority metrics, delivery units can address problems in real time, and experiential learning can occur among civil servants at multiple levels of government.

One well-known example is Malaysia's Performance Management and Delivery Unit, which is credited with improving rice paddy production by working in concert with the Ministry of Agriculture and agricultural extension workers to monitor farmers' cooperatives and help them both to adopt good agricultural practices and participate in in-the-field problem solving.[26] This Malaysian unit has inspired similar models in other countries, including Ethiopia's Agricultural Transformation Agency, Tanzania's former Big Results Now, and South Africa's Operation Phasika.

STRENGTHENING CAPABILITIES AT THE LOCAL LEVEL

Some policies and services necessary for rural revitalization need to be implemented at the local level, especially in countries with more decentralized systems. Thus, strengthening subnational capabilities is equally important. Theoretically, decentralization is expected to improve service delivery by bringing decisions closer to the people who are most directly affected by them, and who can oversee them.[27] In turn, citizens may feel more empowered to participate in decisionmaking and communicate their demands at the local level rather than through national authorities, who may seem distant or anonymous.[28] Local governments can obtain better information about the services needed in their areas of jurisdiction and can be held accountable for their decisions by local voters.

Although evidence to support these claims is mixed, a number of countries have recently pursued the most comprehensive form of decentralization, known as devolution, which involves ceding functions and authority in certain sectors to locally elected officials.[29] Agriculture is one of the most commonly devolved sectors, and the trend toward devolving governance in agriculture has often coincided with major constitutional reforms. For instance, in Pakistan, an amendment to the constitution passed in 2010 stipulated that 17 ministries, including food and agriculture, be devolved to four provinces, resulting in the dissolution of the federal Ministry of Agriculture. In 2010, Kenya's new constitution embraced devolution, and agriculture became the responsibility of the 47 newly created counties. In 2015, Nepal passed a new constitution and transitioned to federalism; agricultural and livestock development are now predominantly provincial powers. In 2014, Zambia initiated the first of a three-phase devolution exercise, resulting in the transfer of responsibility for extension services from the Ministry of Agriculture and Livestock to the country's approximately 110 districts.

However, in these and other countries, the legal transfer of responsibilities for agriculture is not matched by sufficient fiscal decentralization; continuing dependence on the transfer of funds from the central government undermines the autonomy of local policymakers. Where such transfers have been reduced and taxes remain meager, devolution cannot improve service provision as intended. For example, in Ghana, analysis of district budgets over time indicates that total average agricultural expenditures at the subnational level decreased from 11 to 6 percent of spending between the onset of devolution in 2012 and 2015. District governments could no longer rely entirely on the national Ministry of Food and Agriculture for financing, but lacked sufficient tax revenues to make up the difference (Figure 1). Adding to this problem, locally elected politicians tend to allocate budgets to visible public goods that win votes, such as schools and health clinics, rather than to agricultural extension services; this has had negative impacts on the provision of extension services and on government staff morale.[30] The case of Nigeria similarly illustrates the problem of a mismatch between responsibilities and funding; while local governments often have superior agricultural technical expertise, national-level actors play a greater role in determining how the budget for agriculture is allocated—often resulting in suboptimal public investments.[31]

These dynamics reinforce the importance of domestic revenue mobilization to sustain rural service provision and support local government autonomy. Of course, incomes in rural areas tend to be lower than in urban areas, contributing to sharp inequalities in revenue streams and service provision and creating a major barrier to the sustainability of any revitalization efforts. Solutions are being developed. In Tanzania, for example, the use of a mobile money system that allows smallholders to pay taxes on their agricultural produce is currently being piloted to see if it can both provide much-needed resources to local government authorities in rural areas and increase efficiency in tax collection.[32] Other developing countries, including Rwanda and Zambia, are shifting to electronic tax-payment systems for the same purpose.

FIGURE 1 Change in average sectoral distribution of expenditures across metropolitan, municipal, and district assemblies in Ghana

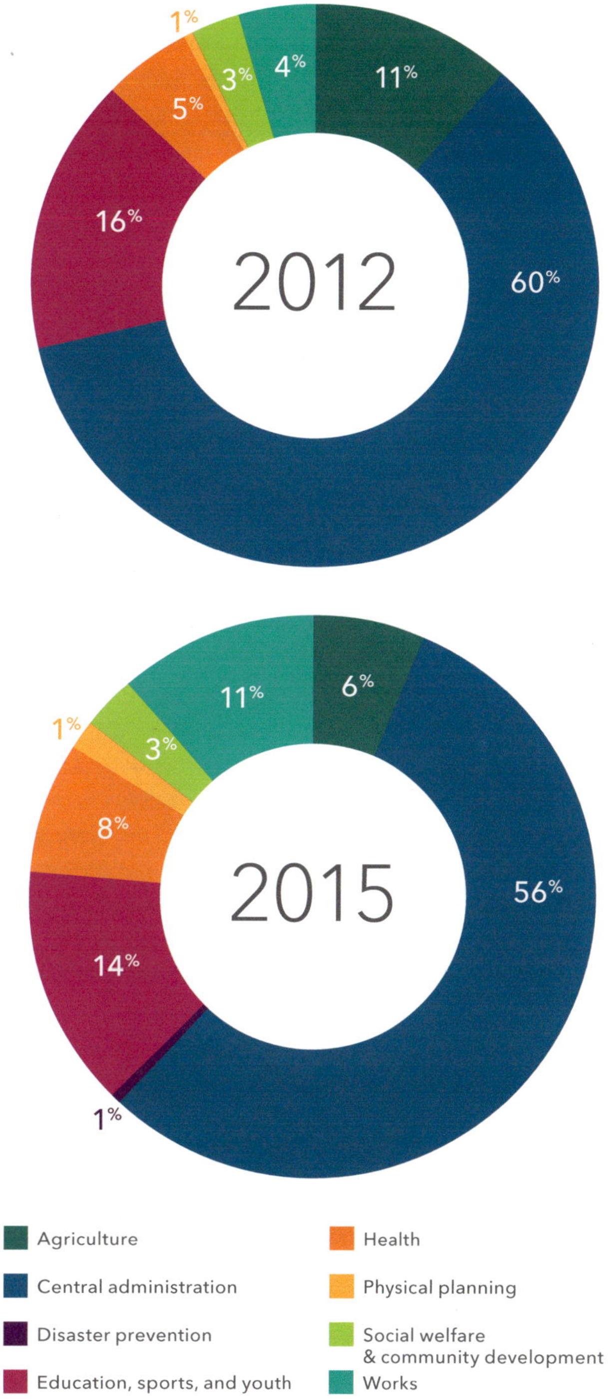

Source: D. Resnick, *The Devolution Revolution: Implications for Agricultural Service Delivery in Ghana*, IFPRI Discussion Paper 1714 (Washington, DC: IFPRI, 2018).

ACCOUNTABILITY TO RURAL CITIZENS

While some of the mechanisms for enhancing state capabilities increase upward accountability from civil servants to their ministerial superiors, mechanisms of downward accountability that allow rural citizens to sanction or reward governments for their performance are equally critical. Downward accountability can facilitate information flows to governments, helping ensure that policies and public investments reflect citizens' real needs and preferences. Among the many mechanisms of downward accountability, we focus here on two key mechanisms: (1) competitive elections and (2) enhanced information and transparency through, among other things, innovations in ICT that have brought about the "information revolution" of the last two decades.

ENSURING COMPETITIVE ELECTIONS

There are many reasons to believe that more competitive elections as well as greater transparency on the part of governments can produce superior public goods and governance.[33] When politicians face a credible threat from opposition parties, they are more inclined to deliver for citizens; otherwise, they risk being voted out of office. Further, when the pool of political candidates grows, it becomes easier for citizens to identify and elect politicians who share their preferences and are likely to represent their interests. Government transparency can help reinforce this electoral accountability mechanism; when citizens are better informed about government mandates, actions, and their consequences, they are in a better position to differentiate good from bad policies and politicians, and thus select politicians who will best serve them. Empirically, this should mean that increases in competition and in transparency improve the quality of governance and public goods provision.

Indeed, evidence from rural settings suggests that more competitive elections contribute to policies and outcomes that benefit rural residents. For instance, the expansion of competitive elections has been associated with higher expenditures on primary education, which is a high policy priority for rural citizens, and with reduced distortionary taxes on rural producers.[34] In Africa, after government disengagement in rural areas in the wake of structural adjustment in the 1980s, growth in electoral accountability in the 2000s resulted in improved agricultural policies, including better-targeted input subsidies that rely on e-payment systems and more streamlined beneficiary lists.[35] These impacts on spending patterns and targeting appear to translate into improved outcomes for citizens. For example, electoral competition in rural areas has been associated with the consumption of more nutrient-dense calories by the poor as well as with increased access to a broad range of amenities and public infrastructure—from improved roads to electricity.[36] Indeed, rural land values per acre and even the prevalence of land rental are higher when political competition is strong.[37]

But political competition is not a panacea that guarantees rural revitalization. Several studies show that political competition can motivate incumbent leaders to create more subnational administrative units, such as districts or provinces, in an attempt to win rural votes and to gain more legislative seats.[38] Yet with already weak revenue bases and inadequately trained bureaucrats, such subdivisions do not necessarily improve service provision for rural citizens.[39] Further, in emerging democracies with low levels of transparency, higher levels of political competition can even lead to worse provision of rural public goods. In such settings, political competition may make it harder for politicians to work together and reach agreements about which goods to supply, where, and how—thus reducing the supply of publicly provided goods.[40] In these environments, policies that increase government transparency, such as making budgets and outlays widely available to the public, and rules that impede partisan redistricting can help ensure that political competition has positive consequences for rural citizens.

HARNESSING THE INFORMATION REVOLUTION

In rural areas, education levels are typically lower than in urban areas, and citizens often have only limited information about the mandates and actions of government. Policymakers often use such information asymmetries to their advantage—and to the detriment of rural citizens' welfare, particularly the welfare of the poorest and least informed.[41] At the same time, the poorest citizens are often the most dependent on government services—such as public education, healthcare, and agricultural extension services—as they cannot afford to buy these services from the private sector. A failure of governmental accountability is therefore likely to harm the poorest citizens most.[42] One possible solution is increasing the information available to the rural poor, thus empowering them to demand accountability.

The task of enhancing transparency is simplified by recent technological innovations and their increased availability to the poor. For example, global mobile phone subscriptions exceeded one per person for the first time in 2016, reflecting a rapid climb over the last 20 years (Figure 2a). This is an important development given that mobile phones have myriad uses, from accessing commodity prices and weather reports to receiving information from the government and contacting politicians. Access to the Internet is similarly on the rise (Figure 2b), offering yet another means of engaging both economically and politically.[43] These trends hold in low-income countries much as they do globally. Indeed, the rate at which high-income countries are adding new Internet users began to slow in the last five years, but the rate in low-income countries accelerated—offering a valuable opportunity for poor countries to catch up.

Just as improving the information available to producers and consumers in the marketplace leads to welfare improvements for both, access to accurate information about what is happening in their country is equally important for empowering rural citizens to make choices about their local and national government and to keep policymakers accountable and responsive.[44] For example, informed voters with access to news media and information about their governments' performance tend to push policymakers to be more responsive to their needs.[45] Innovative tools for disseminating information include one developed by Twaweza, an East African organization, which provides objective data and analysis on government performance using crowdsourced data from mobile phones, and BudgIT, which tracks federal and state-level budgets and expenditures in Nigeria and makes them publicly accessible.

For governments, too, information can be vital for improving the performance of lower-tier governments, bureaucrats, and frontline service providers. For example, an informed central government can benchmark lower-tier government policymakers against

FIGURE 2 Access to technology over time, by country income group

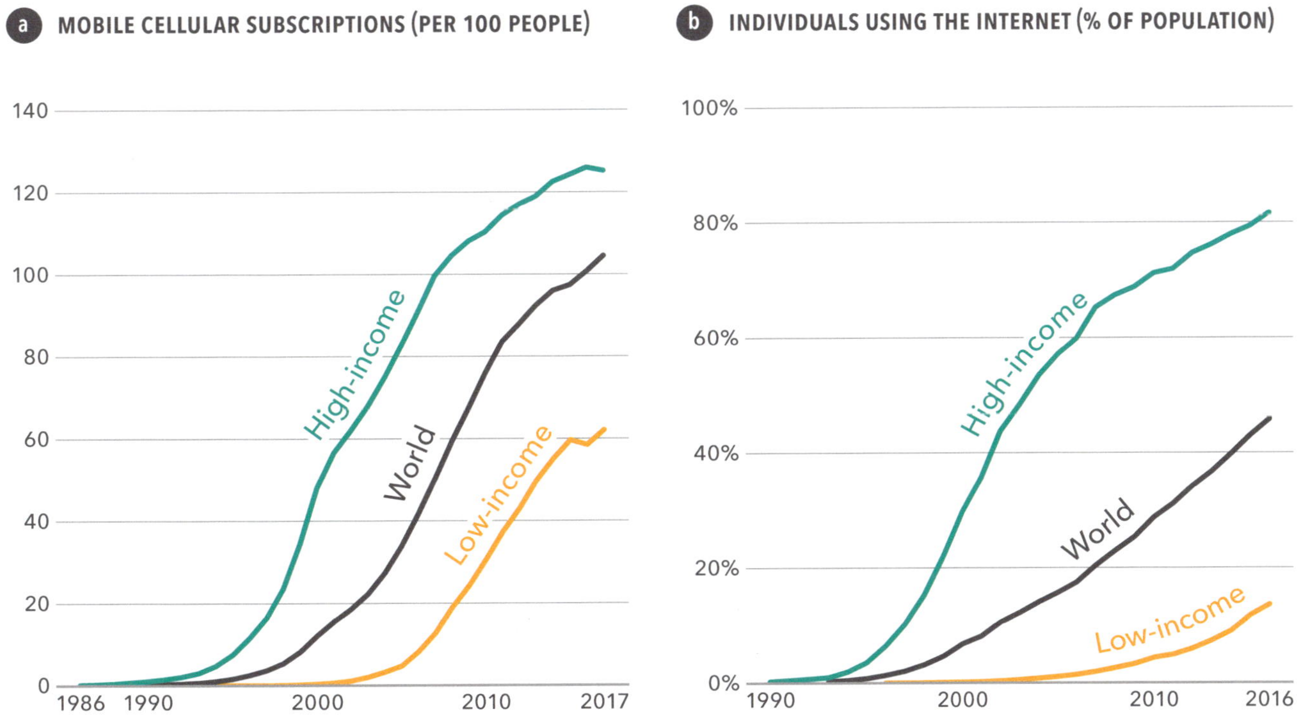

Source: World Bank, World Development Indicators, 2017, https://data.worldbank.org/data-catalog/world-development-indicators.

one another and reward or sanction them based on their relative performance, thus inducing them to exert greater effort.[46] And service providers in many settings are now being required to share their GPS coordinates, take digital photographs, or record videos—with beneficial impacts on their performance.[47] Governments also often use technology to communicate with citizens directly, which has the potential to reduce information asymmetries.[48] Increased access to information thus has the potential for far-reaching impacts on governance in rural areas.

But information alone is not enough. For information to have real impact, recipients must believe it is useful and relevant, and they must have both the power and incentives to act on it.[49] Governments can either help bring this constellation about (for example, by setting up a website or e-governance initiatives, subsidizing access to communication technologies, encouraging citizen dialogue, or otherwise involving citizens in decisionmaking), or they can block the impact of information (for example, by engaging in vote-buying to prevent informed voters from holding policymakers to account).[50] The media, grassroots civic organizations, nongovernmental organizations, and development practitioners can all play a critical role in helping ensure that information achieves its potential for improving government accountability.

IMPROVING GOVERNANCE

Revitalizing rural areas requires that we acknowledge the unique policy complexity of rural settings. Strategies to revitalize rural areas will necessarily be multifaceted, and the specific governance requirements to ensure the success of such strategies will certainly vary across countries and levels of government. Yet as this chapter emphasizes, broadly speaking, strong regulatory regimes, state capabilities, and accountability mechanisms will always provide an important and complementary foundation for tackling poverty, facilitating growth, ensuring high-quality service delivery, and generating jobs. Four broad policy recommendations and tools for improved governance have emerged here:

- Establishing an enabling and predictable regulatory environment is likely to stimulate private sector investment and engagement for rural revitalization. Therefore, it is critical to identify when and why such environments do not exist and ways to better align incentives accordingly. This may require a deeper understanding of the types of relationships between the state and business that have formed over time and why some of these relationships catalyze investment while others do not.
- Building state capabilities is a long-term process, but a number of modalities can encourage bureaucrats to improve service delivery. Performance contracts and delivery units are two possibilities that have now been piloted in dozens of countries and have shown some effectiveness in improving policy implementation at the sectoral and subnational levels—when there is high-level commitment from national ministries, presidents, and prime ministers.
- The recent explosion of access to ICTs provides tremendous new opportunities for citizens, politicians, and bureaucrats at all levels to engage with one another. Given the multiplicity of information sources available, it will be imperative not only to improve access to such innovations but also to ensure that the information they provide is accurate and easy to interpret, especially where education levels are low.
- Rural revitalization involves linkages with secondary towns and cities. However, the rural–urban territorial continuum spans distinct administrative units led by local governments that may have unique development plans, objectives, and constituencies. Establishing intergovernmental coordinating units that facilitate information-sharing across levels of government would be one mechanism to ensure that priorities and resources are aligned in a strategic way to improve desired outcomes, such as better transport connectivity and decent employment.

When considering these and other relevant policy options, it bears emphasizing that while many useful models are available for strengthening governance, from electronic tax payments to digital photographs, there are no quick fixes. This is especially so when there is an overarching lack of political commitment to genuine reform due to ideological resistance, interest group pressures, financial constraints, or all of these. However, recognizing these constraints and identifying their sources is a critical first step toward developing viable and sustainable policy options for rural revitalization.

"Downward accountability can facilitate information flows to governments, helping ensure that policies and public investments reflect citizens' real needs and preferences."

CHAPTER 9

Europe's Experience
Investing in Rural Revitalization

ALAN MATTHEWS

Alan Matthews is professor emeritus of European Agricultural Policy, Trinity College Dublin, Ireland.

KEY FINDINGS

- Rural development is a European Union priority, designated as one of two pillars under the Common Agricultural Policy and supported by almost €100 billion in funding for 2014–2020.
- EU rural development policy aims to foster (1) competitiveness in agriculture; (2) sustainable management of natural resources and climate action; and (3) balanced territorial development of rural economies and communities.
- Spending on rural and farm diversification is expected to create almost 74,000 rural jobs; support for value added and rural services and infrastructure will also improve rural livelihoods.
- With only 11 percent of farm holdings managed by farmers younger than 40, generational renewal efforts support young farmers through start-up grants to enhance their viability.
- Agri-environment-climate measures support improvements in environmental quality, including compensation for farmers who adopt sustainable practices beyond a mandatory baseline.
- "Bottom-up" initiatives emphasize the role of rural communities in determining their own development trajectories.
- Monitoring and evaluation (M&E) is intended to inform policy design, but has fallen short in providing necessary evidence because of data and methodological issues.

KEY RECOMMENDATIONS

Lessons learned from the EU experience support recommendations for rural revitalization efforts in other regions:

- Engage rural areas in protecting and enhancing the natural environment through programs that integrate agriculture with environment and climate objectives.
- Promote endogenous rural development through bottom-up approaches that channel the enthusiasm, skills, and local knowledge of rural communities to develop projects that address the challenges they face.
- Support connectivity of rural areas, particularly through access to the Internet, which is essential to the development of precision agriculture, e-services, and greater rural business innovation.
- Design and implement M&E programs to provide timely evidence on the impact of spending to inform project design and improve targeting and funding allocation.

The countries of the European Union (EU) are highly urbanized and industrialized, but their rural areas remain home to 19 percent of the population and cover 51 percent of the EU's territory.[1] Europe's rural areas are thus a significant source of employment and potential growth, play a vital role in managing environmental assets, and make an important contribution to the recreational options and cultural heritage of European citizens. Yet many of Europe's rural areas are disadvantaged relative to urban areas: Per capita income is lower in rural areas than in other areas—in 2014, it stood at 72 percent of the overall EU average, compared with 121 percent in urban areas. Many rural areas suffer from structural problems, such as a lack of attractive employment opportunities, skill shortages, underinvestment in connectivity and basic services, and a significant youth drain. EU rural development policy is intended to address these issues, to promote rural jobs and growth, and to preserve the region's environmental quality.

EU rural development policy has evolved substantially in recent decades. Now a full-fledged policy at the Union level, its approach to rural revitalization emerged out of agricultural policy and retains a strong farm focus and links to the EU's agricultural programs. The EU rural development policy has three overarching strategic objectives: fostering the competitiveness of agriculture; ensuring the sustainable management of natural resources and climate action; and achieving balanced territorial development of rural economies and communities, including the creation and maintenance of jobs.

Over the last two decades, beginning with fragmented efforts in the 1980s and 1990s and boosted by the 1996 Cork Declaration of the European Conference on Rural Development, rural development has become a key element of EU policy.[2] Since 2000, the EU's Common Agricultural Policy (CAP) has consisted of two pillars: the first is farm income support, and the second is rural development. Rural development policy has evolved over successive CAP programming periods, and spending on rural development has grown relative to traditional agricultural policy spending. Around €99 billion in EU spending—about 24 percent of total EU spending on agricultural and rural policies—is allocated to rural development for the 2014–2020 programming period.

Additional funding for rural development is available through other EU Structural and Investment Funds focused on territorial cohesion policy and fisheries. Moreover, EU member states remain responsible for most expenditures in rural areas, including infrastructure spending, especially on transport and communications; education, health, and social spending; and support for new businesses. This chapter elaborates on EU rural development policy under the second pillar of the CAP.

2014–2020 RURAL DEVELOPMENT PROGRAMMING

EU rural development activities are organized through multiannual Rural Development Programmes (RDPs) that are prepared by member states and subnational regions—member states can choose between having one RDP for the whole country, regional RDPs covering different parts of the territory, or a national program complemented by a set of regional RDPs. For the current programming period, there are 118 RDPs across Europe. The RDPs draw on a menu of permitted interventions set out in EU legislation. Unlike direct payments to farmers funded through the first pillar of the CAP, rural development policies are co-financed by the EU and member states and regions.

Member states and regions base their RDPs on the needs of their territories and aim to address at least four of the following six shared EU priorities:

- Fostering knowledge transfer and innovation in agriculture, forestry, and rural areas
- Enhancing farm viability and competitiveness of all types of agriculture, and promoting innovative farm technologies and the sustainable management of forests
- Promoting food chain organization, animal welfare, and risk management in agriculture
- Restoring, preserving, and enhancing ecosystems related to agriculture and forestry
- Promoting resource efficiency and supporting the shift toward a low-carbon and climate-resilient economy in the agriculture, food, and forestry sectors
- Promoting social inclusion, poverty reduction, and economic development in rural areas

Figure 1 displays the funding shares for the six priorities in the current programming period. More than half of all RDP budgets will be allocated to Priorities 4 (preserving ecosystems) and 5 (resource efficiency and climate resilience); around 15 percent to Priority 6 (social inclusion, poverty alleviation, and economic development—which broadly represent the rural nonfarm sector); 20 percent to Priority 2 (improving agricultural competitiveness—for example, support for investment to modernize farm buildings and machinery); and 10 percent to Priority 3 (promoting food chain organization). Support for Priority 1 (fostering knowledge transfer and innovation) is embedded in the other five through measures such as vocational training, farm advisory services, demonstration farms, and information activities.

Monitoring and evaluation (M&E) constitutes an important element of EU rural development programming, intended to inform policy design through feedback and lessons learned. Evaluations are based on an analytical framework, known as the intervention logic, that provides a systematic description of the

FIGURE 1 Share of funds for EU rural development priorities, 2014–2020

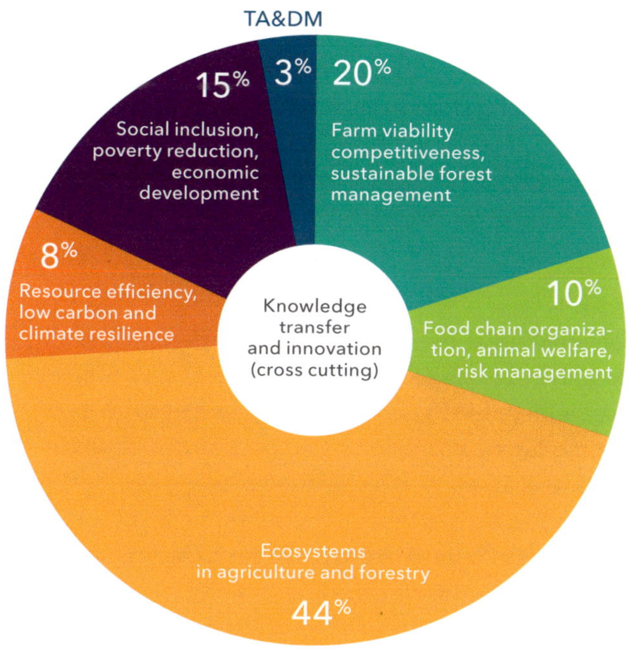

Source: Based on DG AGRI (Directorate-General for Agriculture and Rural Development), *Facts and Figures: Rural Development in the European Union*, prepared for the Cork 2.0 European Conference on Rural Development (Brussels: European Commission, 2016).

Note: TA&DM = Technical assistance and direct management funds spent directly by the European Commission.

links between the actions undertaken and the specific objectives of each measure. A Common Monitoring and Evaluation Framework provides a set of quantitative indicators for conducting M&E.[3] Evaluations are undertaken by independent contractors who report to member states, and the European Commission publishes regular syntheses of their reports.[4]

Despite the resources devoted to M&E, the potential of evaluation to inform future cycles of rural development programming remains unmet.[5] Evaluations often lead to vague or overly general recommendations, and because of lags in data collection, they are not available in time for the subsequent programming cycle. Moreover, they seldom apply counterfactual analysis, often due to lack of data, risking methodologically questionable results. In successive reports, the European Court of Auditors has recommended strengthening the M&E process so that evaluations provide better understanding of the impact of funds spent on rural development.[6]

KEY RURAL DEVELOPMENT MEASURES

Member states and regions can select from among 20 different measures set out in legislation to address their priorities when formulating their RDPs. Examples of measures particularly relevant to broader rural development are highlighted here.

RURAL AND FARM DIVERSIFICATION. One measure supports business startups and provides investment support for nonagricultural activities in order to diversify rural economies and improve rural living conditions. For the 2014–2020 programming period, about 3 percent of planned RDP spending is allocated to these activities, with the expectation of creating 73,800 jobs in supported projects. In Slovenia, for example, the measure supports projects related to adding value to timber, green tourism, treatment of organic waste, and renewable energy sources. Finland, meanwhile, provides innovation vouchers to established small- and medium-sized businesses interested in entering international markets; these vouchers fund the hiring of relevant expertise for developing new products or services. Considerable scope is seen in the potential of the bioeconomy—that is, all sectors based on renewable biological resources, including forestry, aquaculture, renewable energy, and bio-based chemicals—to contribute to rural job creation.[7] Finally, as with all programs subsidizing job-creation activities, care must be taken to avoid creating deadweight (funding projects that would have gone ahead anyway, even in the absence of support) or displacement (where the supported jobs simply replace similar existing jobs, displacing the workers who held them).[8]

ADDING VALUE TO FARM PRODUCTS. Another measure encourages adding value to farm products by taking advantage of growing consumer demand for quality products with a specific provenance. The EU has an extensive scheme of "geographical indications" that protect the names of products whose quality, characteristics, or reputation are linked to their geographical origin. These indications largely apply to food, agricultural products, and wines. Well-known examples include champagne and sherry as well as Parma ham and Stilton cheese. By registering a product with a geographical indication, qualifying producers are entitled to use that indication and are protected against its misuse by producers who cannot meet the criteria. Producers of products with geographical indications can then reap the benefit of any premium that consumers are willing to pay in recognition of the products' quality.

The measure also supports other means of adding value such as on-farm processing, direct marketing, and short supply chains. In Latvia, a group of livestock farmers received grants to buy a mobile slaughterhouse and related equipment. The group set up a nonprofit association to manage the service, and now farmers can choose whether to sell meat products to retailers and slaughterhouses or directly to local markets. In a mountainous part of Lower Silesia in Poland, local businesses received assistance in creating a brand that highlights the unique nature of their products. The "Karkonoska Marka Lokalna–Treasures of the Mountain Spirit" brand unites 29 local producers and service providers and is designed to reflect the heritage and traditions of the region. The branding has helped the various companies involved to increase their sales by promoting their products and services as added-value alternatives to mass-produced products.

SUPPORT FOR COOPERATION. Rural producers are often isolated, with few connections to other producers, and lack bargaining power relative to the businesses that supply their inputs and buy their products. To address these problems, another RDP measure supports cooperation and joint action. Operational groups consisting of multiple actors—such as farmers, researchers, businesses, and nongovernmental organizations—receive support for bottom-up efforts to develop new products, processes, and technologies. For example, strawberry farmers in Germany faced serious yield losses from *Verticilium dahlia*, a disease that infects the soil and causes severe damage to young plants. To find a practical solution, an innovation broker persuaded a research institute, some strawberry farmers, and a small company specializing in the formulation of biological compounds to come together in an operational group funded under a German regional RDP. The project was a success: a biological-control agent was invented and a practical method for application was designed.

Producer organizations for small producers—which concentrate supply, improve the marketing of products, optimize production costs, and carry out research—can also benefit from this measure. Other eligible activities include cooperation among supply chain actors, joint approaches to environmental projects, and local development strategies such as Smart Villages (see below). Chocolate producers in Scotland, for example, have received support for creating a network of artisan chocolatiers to encourage rural tourism.

GENERATIONAL RENEWAL. As in many parts of the world, the average age of farmers in the EU is increasing, with only 11 percent of farm holdings run by young farmers (defined as farmers under 40 years old). The EU aims to reduce the average age of farmers and to support the new generation in adopting sustainable farm practices. Through the first pillar of the CAP, young farmers can receive a top-up payment per hectare to supplement their basic income support. In addition, RDPs may provide lump-sum payments to young farmers, conditional on submission of a business development plan. Member states plan to support almost 180,000 young farmers with this kind of aid during the 2014–2020 period.

These measures help young people who are already set up in farming. For many would-be farmers, however, difficulties in accessing land and credit are major hurdles. More should be done to facilitate the transfer of land and its management from older to younger farmers. Among member states, the ratio of older to younger farmers varies greatly, reflecting different national measures on taxation, farm inheritance, succession planning, and pension entitlements. The European Commission's legal proposal for the CAP post-2020 would require member states to show how they are coordinating these national measures with EU supports to encourage generational renewal.

RURAL SERVICES AND INFRASTRUCTURE. Another measure supports basic services and village renewal in rural areas, including installing broadband infrastructure, setting up or improving local basic services and recreational infrastructure, and restoring cultural and natural heritage. One response to local shops and banks closing in rural areas and public services (schools, health services, police) withdrawing is the creation of multiservice hubs. In Belgium, for example, a village community received support to develop a hub that includes a retail store but also functions as a community center. To increase connectivity in rural areas, the EU trebled the funding available under the RDPs for rural broadband to €0.9 billion over the 2014–2020 period. However, most initiatives to improve rural services and infrastructure are supported by other EU funds or directly by the member states.

ENVIRONMENTAL IMPROVEMENT

EU rural development programs prioritize environmental quality in rural areas, particularly where it is affected by farming activity. For the 2014–2020 period, member states must allocate at least 30 percent of their rural development budgets to environmental and climate action, including both agri-environment-climate measures (called AECMs) and payments to farmers in disadvantaged farming areas (mountainous areas and high-latitude areas).

AECMs complement other elements of the CAP's environmental architecture. These measures include a specific "greening" payment to farmers who follow a prescribed set of farm management practices (maintenance of permanent pasture and conservation areas; crop diversification); the provision of farm extension advice; and regulatory interventions addressing environmental services. The AECM program is voluntary—farmers decide whether they wish to enroll and accept the required management commitments in return for the support provided.

As of 2013, 26 percent of the EU's agricultural area was covered by agri-environmental management commitments.[9] Currently, contracts supporting biodiversity and landscapes cover 13 percent of the EU's farmland area; contracts to improve water management cover 9 percent, as do contracts to prevent soil erosion and improve soil management. In addition, 7 percent of farmland is farmed according to organic principles, for which farms can receive support under RDPs.[10]

For example, the cereal steppes of the Castro Verde region in southern Portugal are one of the last refuges in southern Europe for unique steppe wildlife, including the great bustard, little bustard, and black-bellied sandgrouse. In recent years, changes in land use—such as agricultural intensification, land abandonment, and afforestation—have led to losses of the habitat that steppe wildlife depends on, and many of the most vulnerable species have suffered serious declines. To turn around these losses, the Castro Verde Zonal Program was launched to protect the traditional management of the area and its wildlife. It supports farmers who maintain traditional rotational farming practices and promotes reduced use of pesticides and the control of grazing levels. Since the program began, the types and numbers of both threatened and nonthreatened bird species have increased significantly.[11]

Other measures provide payments to farmers to compensate them for the difficulties of farming in areas with natural constraints, as mentioned above; these payments are intended in part to prevent land abandonment and rural depopulation. Farmers can also be compensated for restrictions they face if their farms are in Natura 2000 areas (a network of valuable habitats designated for environmental protection) or related to river basin management plans. Just over half (54 percent) of EU farmland is designated as facing natural or other constraints.

COMMUNITY-LED LOCAL DEVELOPMENT

The EU's rural development strategy supports "bottom-up" development initiatives through the LEADER approach (an abbreviation of the French *liaisons entre actions de développement de l'économie rurale*, meaning "links among rural development actions"). This approach emphasizes the role of rural communities in determining their own development trajectories based on optimal use of local resources.

Begun as a pilot project in 1991, the LEADER program was mainstreamed into the EU's rural development policy in 2007. LEADER is organized around local action groups consisting of representatives of public and private sectors and civil society. Member states are required to allocate at least 5 percent of their RDP budgets to LEADER activities. For the current programming period, the LEADER approach has been expanded to all EU Structural and Investment Funds as "community-led local development." Around 54 percent of the EU's rural areas are now served by over 3,000 LEADER groups implementing this bottom-up approach to rural development.

Local action groups are chosen in a competitive selection process based on their proposed local development strategy. These strategies identify development needs and local potential, and set out a multiyear plan with measurable objectives. The bulk of their budgets finance private or public-led projects that target entrepreneurs, small-scale infrastructure, social services, training, or other investments. The development of a successful local strategy requires evidence of active community participation and buy-in; local action groups must motivate and enable real participation.

Once funded, the selected groups are responsible for issuing calls for project proposals, selecting the projects that best contribute to the objectives of the local development strategy, and establishing the amount of support. The groups also have a role to play in building the capacity of local actors to develop and implement projects, including fostering their project management capabilities. This enhancement of social capital is both an objective of the LEADER strategy and its means of implementation.

As indicated by its expansion, the LEADER program is generally viewed as a successful model for endogenous rural development. Nonetheless, the approach is not without its critics. An evaluation by the European Court of Auditors noted that LEADER is a complex measure whose added value at local, national, and European levels is not easily assessed. The involvement of local action groups and a multitude of partners increases costs and risks compared with traditional funding methods and should be justified by added value, which member states were not adequately able to demonstrate.[12] The mainstreaming of LEADER has ceded greater control to local administrative authorities, so that fewer community members take part in decisionmaking. In addition, developing an integrated local development strategy as required by the program demands significant human and social capital, so the process may favor more developed and dynamic areas and exacerbate existing spatial disparities in rural development.

Nonetheless, the bottom-up approach to rural development retains wide support. The European Commission and Parliament recently launched the idea of Smart Villages, an initiative to harness digital connectivity for the creation of a better life in rural areas.[13] As with the LEADER approach, Smart Villages are intended to benefit from a cross-sectoral, integrated approach in which rural citizens take the initiative to find practical solutions to the challenges they face and to make the most of new opportunities that are transforming rural areas. Most Smart Village projects will be developed around a single issue—renewable energy production, mobility, or the delivery of e-services such as health and education.

Successful implementation of Smart Villages will depend on overcoming the serious digital gap between rural and urban areas. Only 47 percent of rural EU households have access to fast broadband, compared with more than 80 percent of urban households. Significant funding is available from the European Regional Development Fund and CAP rural development programs to finance the rollout of broadband and other digital infrastructure, especially in remote rural areas.

LESSONS FROM THE EU EXPERIENCE

EU rural regions are diverse, and this diversity became even more marked following the accession of central and eastern European countries to the Union after 2004. Yet these rural regions face common challenges: maintaining competitiveness in an increasingly globalized world; migration patterns that are leading to unbalanced age and gender structures; increasing difficulties in accessing both public and private services in rural areas; and cumulative pressures on the natural environment, most notably farming intensification and its opposite, land abandonment. The EU's development policy supports rural areas in tackling this wide range of economic, environmental, and social challenges. The evolution of this policy and the EU's achievements in these areas, but also where it has fallen short, can provide valuable lessons for other countries pursuing similar rural development objectives.

Although the EU's rural development policy began as an agriculture policy, today rural development is increasingly seen as part of multisectoral, place-based strategies that contribute to the overall objective of territorial cohesion. "Territorial cohesion" is concerned with equity, regardless of location, and aims to develop local potential rather than compensate for disadvantages. Territorial cohesion was incorporated as an EU goal in the Lisbon Treaty of 2007, and is now a legal obligation that must be addressed in EU policies. Rural development policy now must be developed in coordination with other EU funding instruments to achieve efficiencies and synergies. The extension of community-led local development to all EU investment funds is also a step toward greater integration. Nonetheless, in terms of budget allocations (see Figure 1), the policy remains strongly sector-based.

Protecting and enhancing the natural environment is central to the EU rural development policy, accounting for over one-half of the available funds. The basic approach is to compensate farmers for undertaking practices that provide environmental benefits and to promote climate actions that go beyond mandatory minimum standards. Around one-quarter of EU farmland is enrolled in these programs. Since AECMs were first introduced in the mid-1980s, the EU has gained considerable experience in the administration of these schemes, which can be an important resource for other countries facing similar challenges.

An innovative element in EU rural development policy is its pioneering of endogenous rural development driven by local communities. This bottom-up approach, which is now mainstreamed in rural development policy and extended to the other EU investment funds as community-led local development, channels the energy, enthusiasm, skills, and local knowledge of rural communities to develop projects that address the challenges they face and to develop social capital.

Rural areas remain severely handicapped by their lack of high-speed Internet connectivity, but greater funding is being made available to provide the necessary physical infrastructure. Increasing connectivity is essential to achieving the potential of precision agriculture, the delivery of e-services, and greater rural business innovation. In addition, ideas such as Smart Villages are being rolled out to ensure rural communities are able to take advantage of these opportunities.

The EU has placed great emphasis on the importance of M&E as the basis for improved policy design. A common indicator framework has been defined to measure progress; regular evaluations are made of rural development programs; and evaluation methodologies have been improved. But, as successive reports by the European Court of Auditors have pointed out, further progress—in setting clear objectives, identifying the intervention logic behind the measures undertaken, and collecting data to measure results and impacts—is needed.

Looking ahead, the recently published European Commission proposal for the post-2020 CAP, which is currently under debate, announces a major shift in the governance of the CAP. A "new delivery model" will replace the current emphasis on compliance with a focus on results and performance.[14] Moving away from the detailed steering of agricultural and rural policy toward a more enabling approach—with Union-level legislation confined to setting out broad policy guidelines and a menu of interventions—should increase the impact and accountability of the region's rural revitalization effort.

REGIONAL DEVELOPMENTS

AT THE REGIONAL AND COUNTRY LEVEL, DEVELOPMENTS IN 2018 had important repercussions for food security and nutrition. This section offers perspectives on food policy developments across the major regions: Africa, the Middle East and North Africa, Central Asia, South Asia, East and Southeast Asia, and Latin America and the Caribbean. The growing focus on rural revitalization and its potential to spark change is examined for each region, along with many other current topics:

- The African Continental Free Trade Agreement's potential to expand intra-African trade, accelerate integration, and enhance trade competitiveness
- Direct and indirect impacts of continuing conflict on food security and growth in the Middle East and North Africa
- Growing interconnectivity across Central Asia, including transportation, investment, and trade
- Escalating global trade tensions affecting agricultural opportunities in East and Southeast Asia
- Rural transformation and growth of the rural nonfarm sector in South Asia
- Impetus to rural growth from small- and medium-sized towns in Latin America and the Caribbean

AFRICA

OUSMANE BADIANE, JULIA COLLINS, TSITSI MAKOMBE, AND JOHN ULIMWENGU

Ousmane Badiane is the director for Africa, International Food Policy Research Institute (IFPRI), Washington, DC, USA. **Julia Collins** is a research analyst, **Tsitsi Makombe** is a senior program manager, and **John Ulimwengu** is a senior research fellow, Africa Region, IFPRI, Washington, DC, USA.

In 2018, African leaders took important steps to accelerate agricultural and rural transformation across the continent. These include the January launch of the first-ever continental Biennial Review of countries' progress in meeting commitments to agriculture sector development. In addition, the signing of the African Continental Free Trade Agreement in March moved the continent toward tighter integration and greater intraregional trade through the free movement of goods and people. Progress in reducing poverty and hunger has continued, but levels remain high, particularly in rural areas. While there is a pressing need to provide for the poor, governments should target investments to areas of synergy between improving social services and promoting productivity growth to ensure that recent increases in agricultural productivity are sustained.

TRENDS IN GROWTH AND RURAL DEVELOPMENT

After a slowdown in 2016 due in part to lower commodity prices and weaker global growth, gross domestic product (GDP) growth in Africa south of the Sahara rose to 2.6 percent in 2017 and was expected to reach 3.1 percent in 2018.[1] Higher commodity prices have helped exporting countries to improve their fiscal situation, but many African economies remain vulnerable to global risks, and growth remains weaker than in the 2000s. Poverty, as measured by the international poverty line of $1.90 per day, continued to fall, declining at around 1 percent per year between 2003 and 2017.[2] However, 36 percent of the population of Africa as a whole remained poor as of 2017. An acceleration in poverty reduction is urgently needed. Similarly, hunger indicators are improving, but levels remain high. In Africa as a whole, 17.6 percent of the population was undernourished in 2015, and among children under five, stunting (low height-for-age) affected 33.3 percent and underweight (low weight-for-age) affected 18.5 percent in 2017.

Agricultural value added for Africa grew at an average annual rate of about 4.3 percent during the period 2003 to 2017, meaning that the continent did not meet the target of 6 percent annual agricultural growth set under the Comprehensive Africa Agriculture Development Programme (CAADP). However, West Africa did meet the 6 percent target during the 2003–2008 period, and North Africa exceeded the target during the 2008–2017 period. Agricultural labor productivity grew at 1.6 percent annually from 2003 to 2008 and at 2.5 percent annually from 2008 to 2017, while land productivity grew more rapidly, at 3.2 and 5.3 percent during the two periods, respectively (Figure 1). These productivity growth rates are the highest in several decades, and reflect reforms enacted in previous decades to improve the environment for agriculture and private sector participation, as well as increased public investments in agriculture in the years following CAADP's launch in 2003. Continued productivity growth suggests that agriculture is contributing positively to reducing poverty and hunger and to increasing economywide productivity.[3]

The share of agriculture in total public expenditures averaged 3.5 percent during 2003–2008 and 3.0 percent during 2008–2017, falling short of the CAADP goal of a 10 percent agricultural public expenditure share (Figure 1). Five countries (Burkina Faso, Ethiopia, Madagascar, Malawi, and Mali) met the 10 percent goal during the 2008–2017 period,

FIGURE 1 Change in poverty and agriculture indicators for Africa

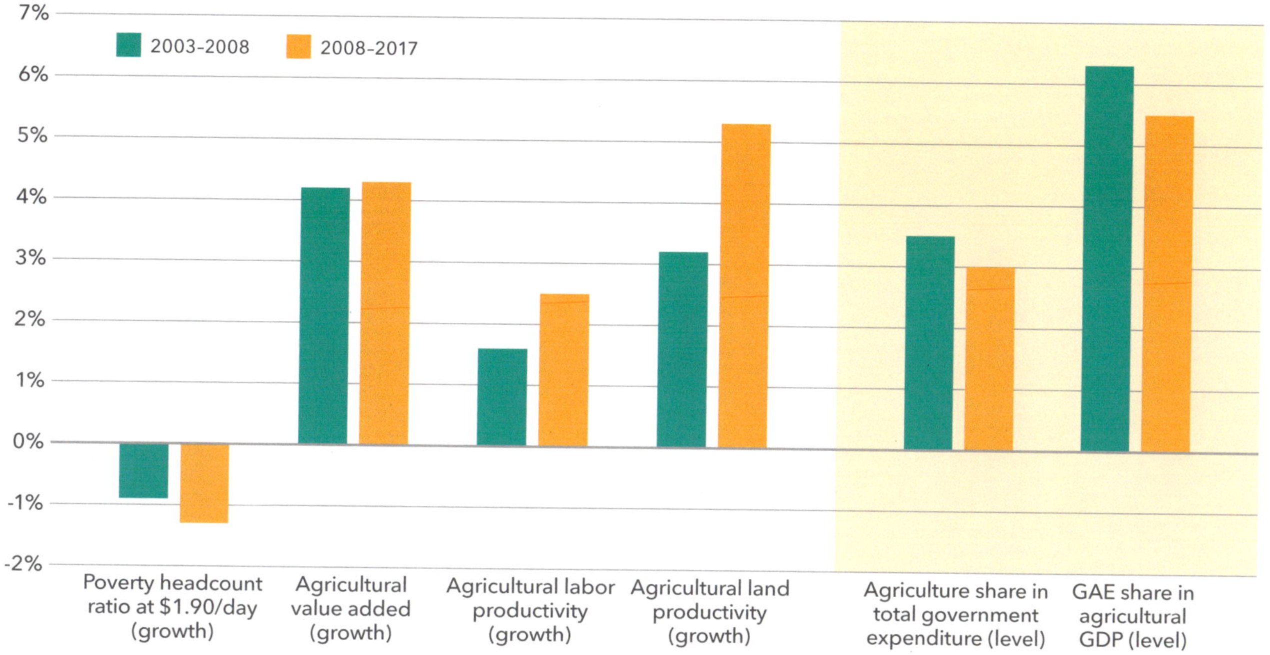

Source: T. Makombe, W. Tefera, and S. Benin, "Tracking Key CAADP Indicators and Implementation Processes," in *Boosting Growth to End Hunger by 2025: The Role of Social Protection, 2017–2018 ReSAKSS Annual Trends and Outlook Report*, ed. F. Wouterse and A. S. Taffesse (Washington, DC: IFPRI, 2018).

Note: GAE = government agricultural expenditure. GDP = gross domestic product.

and Senegal came close to reaching the goal. In Africa as a whole, the share of public agricultural expenditures in agricultural GDP also decreased, from 6.3 percent in 2003–2008 to 5.5 percent in 2008–2017. Although absolute levels of public expenditures in agriculture rose between 2003 and 2008, expenditures actually decreased between 2008 and 2017. At the same time, expenditures for social services such as health, education, and social protection rose sharply. This reflects the trade-offs that governments face due to the pressing needs of the poor; however, reversing the decline in agricultural expenditures is imperative if we are to spur continued productivity growth, improve livelihoods, and accelerate poverty reduction.

Rural areas tend to have significantly higher rates of poverty and hunger. For example, analysis of household survey data suggests that in 2015, poverty rates in rural Kenya were 11 percentage points higher than in urban areas, while the gap between rural and urban poverty rates was over 20 percentage points in Côte d'Ivoire.[4] Hunger rates, as proxied by the share of the population below the national food poverty line, were around 10 percentage points higher in rural areas in both Kenya and Côte d'Ivoire. In both countries, the poverty gap has narrowed since the 2000s, due to greater declines in rural poverty in Kenya and increases in urban poverty in Côte d'Ivoire. However, in other countries, including Malawi and Burkina Faso, the gap widened as poverty fell more rapidly in urban areas than in rural areas (Figure 2).[5] Rural areas also have significantly less access to services than urban areas. As of 2016, only 24.8 percent of the rural population in Africa south of the Sahara had access to electricity versus 75.7 percent of the urban population. And urban residents were twice as likely as rural residents to have access to basic drinking water and sanitation services.[6] The combination of high poverty rates and poor services drives high levels of hunger and undernutrition in rural areas.

FIGURE 2 Change in rural and urban poverty rates in select African countries

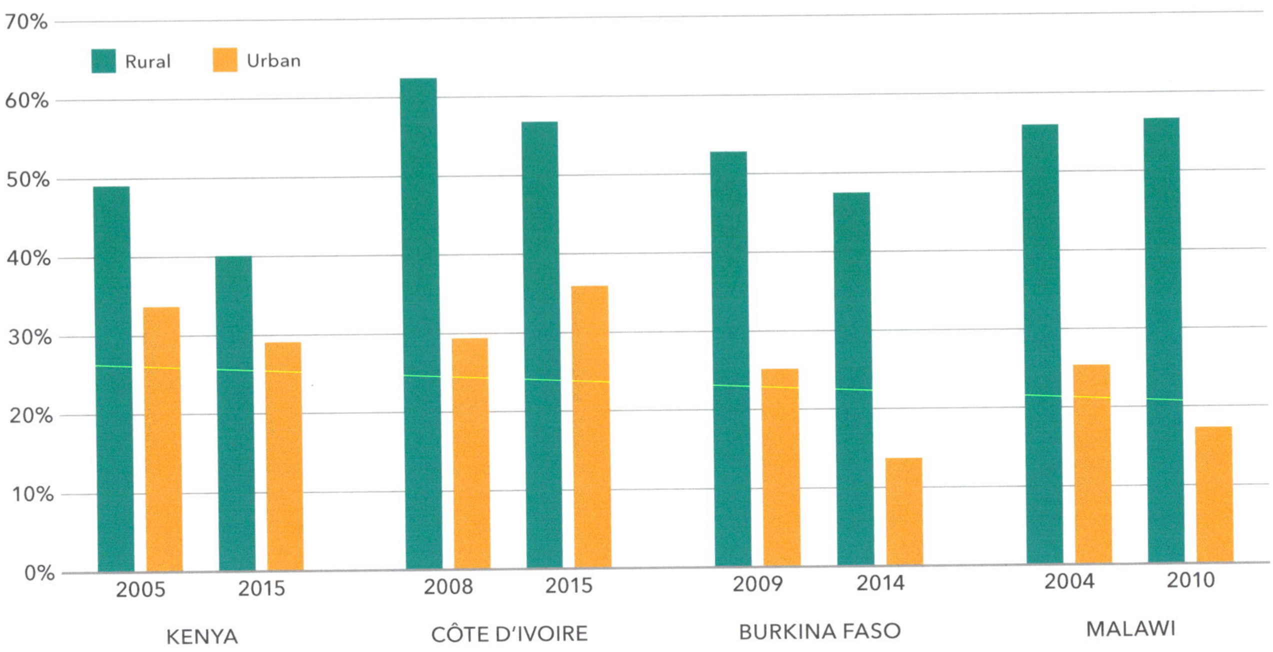

Source: Authors, based on the World Bank's World Development Indicators database, accessed November 2018, and IFPRI microsimulation results.

Note: Poverty is measured using national poverty lines, which are not comparable across countries. The 2015 data for Kenya are estimated.

POSITIVE POLITICAL DEVELOPMENTS AND POTENTIAL ECONOMIC RISKS

While conflicts in Nigeria, Somalia, and the Central African Republic are unresolved, several recent political developments bode well for greater peace, security, and economic stability. Ethiopia and Eritrea signed a peace agreement in July 2018, ending a two-decade-long cold war and reopening trade ties.[7] In early 2018, political changes in South Africa gave rise to hopes for reduced corruption and political renewal. In South Sudan, leaders signed a peace agreement in September, potentially ending a civil war that has displaced millions and killed tens of thousands since 2013.[8]

However, serious threats to growth and food security remain. Foreign direct investment inflows to Africa in 2017 dropped by 21 percent from their 2016 levels, with much of the decline attributable to still-low oil and other commodity prices.[9] The ramping up of global trade wars could be detrimental to growth and welfare worldwide by raising uncertainties around investment, creating barriers to the sharing of technology, and reducing the purchasing power of consumers.[10] Some African analysts foresee potential opportunities in changing markets, such as the possibility of increasing African exports to the United States, as well as potential threats, including reduced capacity of domestic producers to compete with Chinese imports in African markets.[11] Many countries are still struggling with high levels of debt stemming in part from commodity price declines in the past several years; in 2017, 15 countries in Africa south of the Sahara were deemed to be in debt distress or at high risk of debt distress.[12]

TRADE POLICY DEVELOPMENTS AND RURAL ECONOMIES

In 2018, African countries invested in institutional support for agricultural and rural transformation with the launch of the inaugural Africa Agriculture Transformation Scorecard (AATS) and the CAADP Biennial Review report. Their launch during the African Union (AU) Summit in January 2018 marked a milestone in promoting mutual accountability at the highest political level and in strengthening evidence-based agricultural planning and implementation, as called for in the 2014 Malabo Declaration. With 47 of 55 AU member states submitting reports, the inaugural Biennial Review was a major success.

The signing of the African Continental Free Trade Agreement by 49 African countries as of July 2018 was another major milestone. This agreement is an ambitious endeavor that aims to create a single African market for goods and services, along with free movement of businesspeople and investments, and to establish a continental customs union.[13] These measures are expected to expand intra-African trade (which remains low compared with intraregional trade in other world regions), accelerate regional and continental integration, and enhance trade competitiveness.[14] To enter into force, the agreement must be ratified by at least 22 countries; as of late 2018, 18 countries had ratified it.[15] The AU carried out an advocacy campaign to urge more countries to ratify ahead of the January 2019 AU Summit. And although Africa's economic powerhouses, South Africa and Nigeria, initially did not sign the agreement, South Africa signed in July and Nigeria's president signaled that his country would also sign.[16] These developments may persuade more countries to ratify the agreement soon. While implementation, monitoring, and enforcement of the Africa-wide agreement will likely be complex, if successfully implemented, the African Continental Free Trade Area would be one of the world's largest free trade markets, with 1.2 billion people across 55 countries and a GDP of US$2.2 trillion.[17]

The 2016 slowdown in Africa's economic growth exposed the continent's continued economic vulnerability and underscored the need to diversify Africa's export base and raise productivity and value added in the agriculture sector. The African Continental Free Trade Agreement presents Africa with an opportunity not only to improve productivity and value added through trade, but also to create decent jobs, raise incomes, and reduce economic vulnerability. Intra-African trade is expected to grow by 53 percent solely as a result of the elimination of import duties.[18] Trade facilitation efforts under the agreement will help to boost trade and also promote trade diversification, economic growth, and industrialization, in particular of the agribusiness sector.

POLICIES FOR RURAL REVITALIZATION

Africa is urbanizing rapidly, creating new opportunities for rural transformation and revitalization, in large part through growing demand for food. Patterns of urbanization in Africa differ from those in other regions, with relatively more growth in small and medium cities than in large cities.[19] The strong growth in smaller cities and towns presents opportunities for rural producers and agribusiness firms to cater to urban demand for more convenient processed forms of local foods. Retail inventories conducted in Ghana, Mali, and Tanzania have documented the significant presence of domestic brands of processed local dairy and grain products.[20] In Senegal, following years of low and declining millet consumption in Dakar and other cities, new processing technologies have fueled a rapid expansion in small firms producing ready-to-cook and ready-to-eat millet products that have boosted urban consumption.[21] As a result, employment opportunities in off-farm segments of the agrifood system are growing rapidly and are projected to continue rising.[22] However, firms must adapt to meet consumer preferences for food safety, quality, and marketing to capture a greater share of urban markets. Focus group discussions in Ghana and Nigeria have found that urban consumers prefer traditional foods, but often lack confidence in the quality of locally processed food products.[23]

POLICIES TO PROMOTE RURAL ENTERPRISE GROWTH

Rural areas will benefit greatly from strategies to promote private sector investment in the emerging agro-processing sector. Growth in the agribusiness sector is likely to be particularly effective in reducing poverty, given the sector's geographic dispersion and strong linkages to upstream and downstream value chain actors.[24] Policies to boost enterprise growth should aim to strengthen markets to lower transaction costs; increase access to management-skills training to enable innovation; facilitate the transfer and adaptation of knowledge and technologies from abroad; and increase access to credit.[25] Several African countries have invested in comprehensive, place-based strategies to increase private investment in agriculture, sometimes referred to as agribusiness parks or staple crop processing zones. For example, the Democratic Republic of the Congo established a special agricultural economic zone in 2014, incorporating public-private partnership companies as well as smallholder farmers. Actors including the government, private sector, development organizations, and farmer organizations offer smallholders mechanization and extension services, infrastructure improvements, access to inputs, and output purchase agreements. Smallholders involved in the initiative have increased both the land area cultivated and cassava yields.[26] Such place-based strategies can contribute to rural transformation and sustainable development in Africa by encouraging innovation through technology transfer along agricultural value chains; providing opportunities to introduce reforms at a small scale; and bringing together multiple actors to carry out complementary actions along the value chain.

LINKING RURAL AREAS TO GROWING URBAN DEMAND

Better infrastructure is also urgently needed to enable rural producers and firms to seize opportunities presented by increased urban demand. Infrastructure in Africa tends be of poorer quality and yet more expensive than in other regions, creating bottlenecks in potential food value chains.[27] Poor rural roads, for example, affect the ability of processing firms to obtain farm products from local producers reliably and on time; and inconsistent access to electricity raises firms' costs.[28] Renewable energy is increasingly affordable, but much needs to be done to take advantage of recent advances in order to bring power to rural communities. However, Africa is more advanced in terms of mobile connectivity, and closer to connectivity levels in the rest of the world.[29] Further progress in expanding mobile Internet would enable broader access to content, such as technical education and skills development. Public investments in the transport and communications sectors in Africa south of the Sahara increased over the 2000s, but investment levels remain insufficient to meet the region's infrastructure needs.[30] The Programme for Infrastructure Development in Africa (PIDA), an initiative of the AU Commission, NEPAD Planning and Coordinating Agency, and African Development Bank, identifies strategies and priorities and facilitates investments in a variety of infrastructure projects in Africa.[31] In November 2018, the African Development Bank approved a US$40 million investment in the African Infrastructure Investment Fund, which will support investments in transport, energy, and telecommunications infrastructure.[32] Private investments in solar power have surged in recent years, enabling the expansion of electricity to hundreds of thousands of customers who are not connected to existing electrical grids.[33]

BETTER WATER TECHNOLOGIES

In the face of a changing climate with more frequent weather extremes, the longstanding need to invest in water management in rural Africa is increasingly urgent. Many of the poorest and most vulnerable areas of Africa are those where water is the main constraint to agricultural productivity. Increasing water availability is thus an indispensable component of strategies to reduce poverty. Only 6 percent of agricultural land in Africa is irrigated, a much lower share than in other world regions. However, Africa has high potential for expanding irrigation—an IFPRI study estimates that irrigation could be expanded profitably to 24 million hectares by 2050, from the current 13 million, with most of this expansion potential in Africa south of the Sahara.[34] Increasing irrigation would bring multiple benefits, including a longer growing season and mitigation of weather risks, with higher income opportunities for smallholders; the possibility of producing more diverse and nutritious crops; and increased rural employment opportunities. Several African countries have achieved rapid expansion of irrigated areas since the early 2000s, including Ethiopia, Kenya, and Niger. Strategies in these countries have included improved water and irrigation regulations, stronger partnerships with the private sector, the use of new technologies, and in some cases a prioritization of small-scale irrigation.

SOCIAL SERVICES FOR INCREASED PRODUCTIVITY AND RESILIENCE

Finally, strategies to increase rural growth must also make rural areas more livable by expanding access to social services, including healthcare, education, and social safety nets. Social protection programs such as Ethiopia's Protective Safety Net Programme (PSNP) have successfully increased the food security and assets of the country's poorest and most vulnerable rural residents.[35] Numerous studies have demonstrated the importance of education and healthcare to increasing agricultural productivity.[36] Improved social services in rural areas, particularly education and training, will also be key to helping ensure that agriculture remains an attractive sector for Africa's growing youth population. In particular, vocational training is an important component of strategies to prepare youth for current and future employment opportunities arising from transforming value chains, increased agro-processing, and on-farm technological changes such as increasing mechanization. Investments in services in rural areas today will increase productivity and growth into the future, both in the agriculture sector and in the wider economy.

OUTLOOK FOR 2019

In 2019, GDP growth is projected to accelerate to 3.8 percent in Africa south of the Sahara.[37] The African Continental Free Trade Agreement is expected to come into force, advancing regional trade and integration and contributing to improvements in competitiveness and employment. African leaders will build on recent advances in mutual accountability in the agriculture sector by preparing the second CAADP Biennial Review, to be launched at the AU Summit in January 2020. Consolidation of peace processes in Ethiopia, Eritrea, and South Sudan, if sustained, will open the door to faster growth and poverty reduction in these countries. However, high levels of debt will continue to pose a challenge for African countries as this debt reduces the fiscal space for necessary investments in agriculture, infrastructure, and social services. In addition, global trade wars threaten to reduce growth and investments worldwide. Despite these constraints, African countries must continue to make productivity-enhancing investments and provide a supportive environment for farmers and agribusinesses across the value chain to respond to growing opportunities in the agriculture sector.

MIDDLE EAST AND NORTH AFRICA

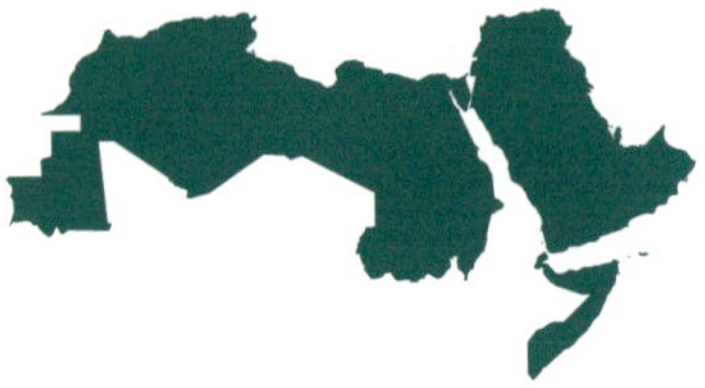

CLEMENS BREISINGER, NADIM KHOURI, AND FATMA MAHFOUZ

Clemens Breisinger is a senior research fellow and leader of the Egypt Country Strategy Support Program, International Food Policy Research Institute (IFPRI), Cairo, Egypt. **Nadim Khouri** is an independent researcher supporting the Global Agriculture and Food Security Program and the Egypt Country Strategy Support Program. **Fatma Mahfouz** is a research associate, Development Strategy and Governance Division, IFPRI, Cairo, Egypt.

Relatively stable world food prices in 2018 were a boon for all countries of the Middle East and North Africa (MENA), which, as net food importers, saw improvements in the region's food affordability index (Table 1). Global fuel prices also increased, but at a faster pace than food prices.[1] The higher fuel prices reduced the pressure for economic reform in oil-exporting countries but increased the pressure for reform in the oil-importing countries, especially for reform of energy subsidies. Oil continues to shape development in the MENA region—along with issues related to finding employment for the region's expanding youth population; arid, natural-resource challenged rural environments; and conflict.

THE TWO MENAS: CONFLICT AND STABILITY

Conflict-related food insecurity in the region continued to increase, reflecting infrastructure damage (Yemen, Iraq, Syria, Libya), supply chain disruptions (Libya, Iraq), and soaring input prices (Iraq, Yemen, Syria).[2] However, there is reason for cautious optimism in Syria at the time of writing this report (November 2018) as the country's major battles, with the exception of Idlib, have ended.[3] Yemen "remains the world's largest humanitarian disaster with an ongoing fight against famine, civilians dying from preventable diseases and an economy on the verge of collapse," in the words of a UN special envoy.[4] Fighting intensified in many areas, particularly in and around Hodeida, a port that is vital for food imports and the survival of Yemenis, as only a fraction of the country's food needs can be grown domestically, even in peaceful times. As a result, more than half of the country's population of 29 million is food insecure—with 8 million relying entirely on external food assistance.[5] Other flashpoints of violence in MENA include Iraq, Libya, and the West Bank and Gaza.

MENA countries that are indirectly affected by conflict, especially Lebanon and Jordan, continue to struggle with managing their large refugee populations. In some cases, tensions are building between refugees and host governments and populations. In Lebanon, the refugee situation, compounded by the weakness of Lebanon's economic and political institutions, continues to constrain investments, tourism, and trade; Lebanon's real annual gross domestic product (GDP) growth is projected to remain modest at 2.7 percent between 2018 and 2022.[6] Similarly, in Jordan, which borders two conflict states (Iraq and Syria), growth recovery has been slow. Jordan's real annual GDP growth is projected to average just 2.6 percent from 2018 to 2020, and protests over austerity measures are impeding further structural reforms.[7]

The economic outlook for Egypt improved in 2018. As a result of growing confidence in the country's stability and commitment to the ongoing economic reform process, growth is projected to reach 5.5 percent in 2019, up from 4.2 percent in 2017.[8] But growth rates in other MENA countries were more modest—11 out of 21 MENA countries reported negative per capita growth rates between 2016 and 2017 (Table 1). For oil exporters, the record low international oil prices from 2015 through 2017 largely explain the current negative per capita growth rates. In response, most of the oil exporters (Bahrain, Qatar, Saudi Arabia, and the United Arab Emirates) adopted business-friendly policies in 2017 and 2018 to attract foreign investment. Oman and Kuwait also pushed forward with policies to reduce the number of expatriates as a means to increase the employment of nationals. Such reforms that are likely to harm the MENA countries that benefit from workers' remittances.[9]

TABLE 1 Overview of key indicators and policy changes in MENA

	GDP per capita annual % change 2016–2017, or latest two years[a]	Affordability of food (index)[b]	Policy changes[c] (2016/2017/2018) 🟠 Policy reform in 2016 ⚫ Policy reform in 2017 🟣 Policy reform in 2018			
			Macro/ fiscal/trade	Investment/ investment climate	Social protection	Agriculture
Oil exporters						
Algeria	-0.1	-0.1	🟠⚫⚫⚫🟣	🟠		⚫
Bahrain	-0.8	-0.2	🟠🟠⚫⚫⚫🟣	🟣🟣		
Kuwait	-4.8	-0.1	⚫🟣			
Oman	-4.8	-0.3	🟠⚫⚫		🟣	
Qatar	-1.1	-0.3	🟠⚫	🟠🟣🟣		
Saudi Arabia	-2.7	0.8	🟠🟠⚫⚫⚫⚫🟣🟣	🟠🟣	🟣	🟠⚫🟣
United Arab Emirates	-0.6	0.2	⚫🟣🟣🟣	🟠🟠⚫🟣🟣🟣		
Oil exporters affected by conflict						
Iraq	-3.6	–	🟠🟠🟣		⚫🟣	
Libya	25.1	–			⚫🟣	
Sudan	1.8	-0.1	🟠⚫🟣			
Yemen	-35.9	-0.2				
Oil importers						
Comoros	0.2	–	⚫			
Djibouti	2.5	–	🟠🟠			🟠
Egypt	2.2	0	🟠🟠🟠⚫⚫⚫🟣🟣🟣	🟠⚫🟣	⚫⚫🟣	
Jordan	-0.6	-0.1	🟠🟠⚫🟣	🟠⚫🟣🟣	🟣	🟠
Lebanon	0.8	–	🟠🟣🟣		🟠🟣	
Mauritania	0.7	–	🟣	⚫		🟣
Morocco	2.7	0	🟣🟣	🟠🟠🟣	⚫🟣	
Tunisia	0.8	0.3	🟠	⚫	⚫	
West Bank and Gaza	-6.1	–				
Oil importers affected by conflict						
Syria	–	-0.1		🟣🟣	🟠🟣	
Somalia	–	–			⚫	

Source: [a] **GDP per capita figures** are from World Bank, World Development Indicators Database, accessed September 2018. [b] **Affordability of food figures** are based on Economist Intelligence Unit, Food Security Index. [c] **Policy changes** are based on authors' compilation from Economist Intelligence Unit, Country Reports (2016, 2017, and 2018).

Note: – indicates no data available. GDP = gross domestic product. Affordability includes the following indicators: food consumption as a share of household expenditure, proportion of population under global poverty line, GDP per capita (US$ PPP), agricultural import tariffs, presence of food safety net programs, and access to financing for farmers.

BEYOND MACROECONOMIC REFORMS: ADDRESSING UNEMPLOYMENT

With low and in some cases even negative growth rates, MENA economies cannot meet the urgent need for jobs for the region's growing population. Total youth unemployment (between the ages of 15 and 29) is more than 27.5 percent in the Middle East and 30.5 percent in North Africa.[10] Oil-importing countries have undertaken some initiatives to deal with growing unemployment: directly through improvements in the business environment designed to boost private sector-led growth and achieve an even distribution of employment opportunities across the urban-rural divide (Egypt, Jordan, Mauritania); and indirectly through reform of labor markets and the education system (Morocco). The Syrian Investment Authority, in an ambitious attempt to mitigate the impact of conflict, drafted tax incentives to encourage import-substitution and export-oriented projects.[11] However, for the future needs of the region, doubts are being raised as to whether the traditional industry-based, export-oriented model of growth will work (as it did for many Asian countries)—and how a technology-driven, digital "new economy" could help absorb the excess labor force.[12] The impact of the digital economy on rural areas in MENA may be particularly beneficial, as it could afford new prospects for youth to revitalize rural areas; the aging population of rural areas is unlikely to manage the agricultural and food system transformation. And youth involvement in the rural economy could slow migration to MENA's urban areas, where unemployment is the highest in the world. Young women and men are attracted to better education in urban areas and are willing to "wait" for the few jobs there rather than move to rural areas, thus contributing to an aging of MENA's rural areas and unemployment in urban areas.[13]

Table 1 summarizes the region's key policy changes between 2016 and 2018. These macroeconomic reforms and various efforts to improve the business climate are necessary, but likely insufficient to create urgently needed jobs and reduce poverty and food insecurity. Additional sectoral or area-specific policies and investments are required. Given the region's large rural populations and the potential of rural areas, greater attention should be given to fostering rural development as a driver of economywide development.[14]

A FRESH APPROACH TO RURAL DEVELOPMENT

Despite agricultural support policies that often favor rural areas in MENA countries, significant inequality persists between rural and urban areas (Figure 1). Globally, a pattern of diverging rural–urban living standards is common as countries become more urbanized, until they reach upper-middle-income levels.[15] Consistent with this global pattern, high- and upper-middle-income countries in MENA tend to have less rural–urban inequality than do lower-middle-income countries. Because urban areas generally enjoy more favorable geography, infrastructure, and human capital, they attract more private sector firms, investment, and job opportunities. But there are good reasons for governments in MENA to rethink their agricultural support policies and broaden their policy scope toward revitalizing rural areas, especially in upper- and lower-middle-income countries. Rural areas are still home to 25 percent of people in upper-middle-income countries and 51 percent in lower-middle-income countries, and job growth in urban areas is too slow to absorb the growing number of rural workers.[16] Moreover, rural economies of the 21st century can reap the benefits of technological advances that allow for the modernization and diversification of traditional rural sectors such as agriculture, agro-processing, and tourism and attract new investments into nontraditional rural sectors that emerge with the spread of digital technology.

Adopting a rural revitalization approach will require assessing MENA's institutions, policies, and public investments in terms of their effectiveness in fostering rural development. For example, does the current, often centralized institutional setup incentivize local governments to advance local development? Does current agricultural policy, which often favors cereals over fruits and vegetables, contribute to rural revitalization? Is the use of scarce public resources for food and fertilizer subsidies the most effective way to support growth in rural areas in the 21st century? Beyond agriculture, what other sectors could drive growth in rural areas, and how can rural economies be better linked to urban economies, for example through food supply chains?

MODELS FOR RURAL REVITALIZATION

Outside of MENA's major cities, regional economies still depend heavily on traditional rural and on-farm economic activities, and the need to move toward more diverse and modern rural employment is widely recognized.[17]

Global experience suggests that modernizing rural economies requires shifting decisionmaking power to local governments and local communities. Such was the Chinese experience, where the close collaboration between central and local governments proved to be essential for the success

FIGURE 1 Rural share of population and rural-urban inequality

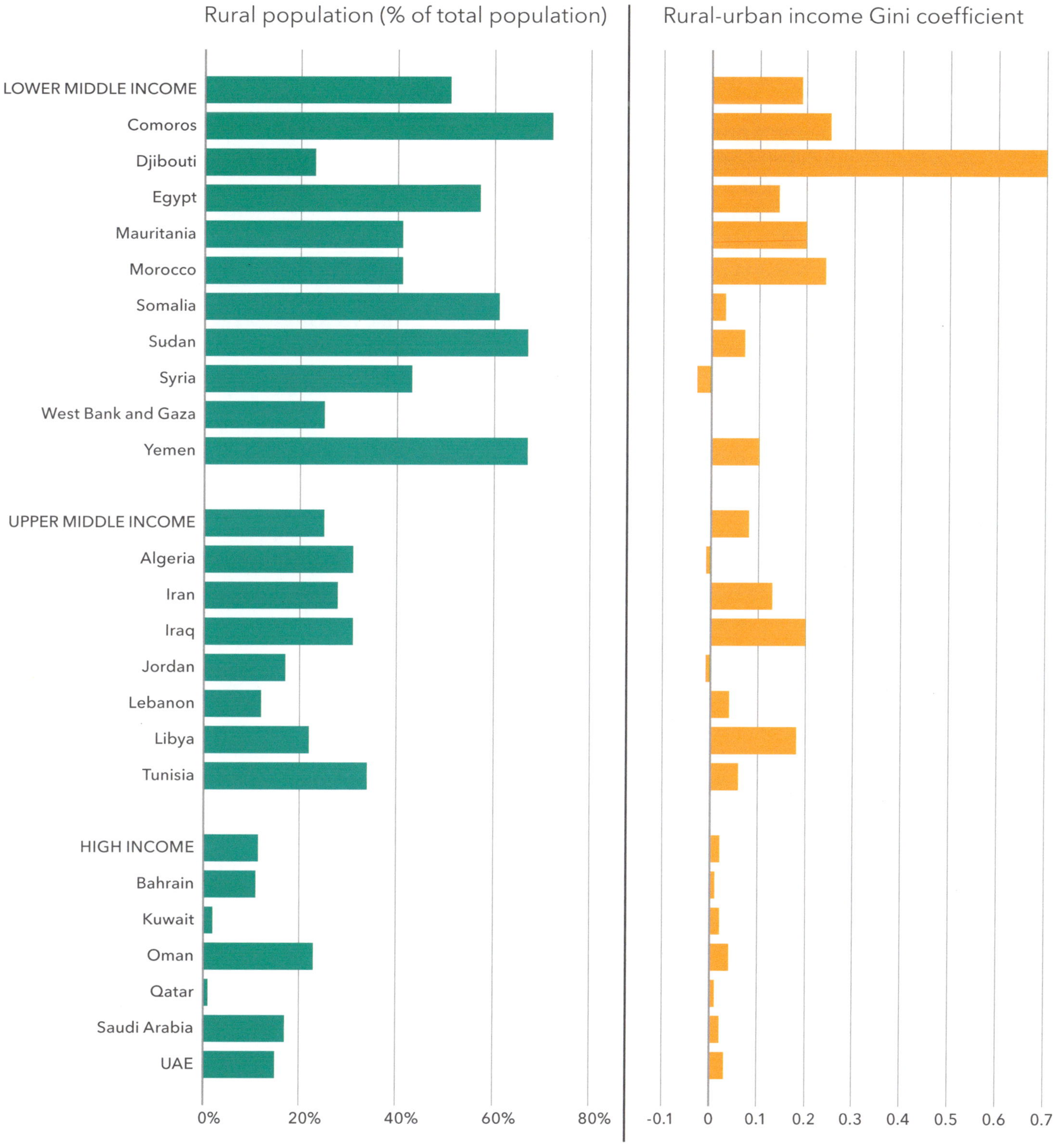

Source: A. Nin Pratt et al., *Agriculture and Economic Transformation in the Middle East and North Africa: A Review of the Past with Lessons for the Future*, Food Policy Report (Washington, DC, and Rome: IFPRI and FAO, 2018), based on World Bank, 2016 World Development Indicators , http://databank.worldbank.org.

Note: The rural-urban income Gini coefficient is the share of agricultural GDP minus the share of agriculture in employment, which provides a measure of the income gap between the rural and urban sectors. See C. P. Timmer, "The Structural Transformation and the Changing Role of Agriculture in Economic Development: Empirics and Implications," Wendt Lecture, October 30, 2007, American Enterprise Institute, Washington, DC.

of many of its industrial parks. In these parks, the central government focused on providing infrastructure to connect rural and urban economies, a business-friendly environment, and public services, and local governments worked closely with local communities and businesses to address common bottlenecks and promote private investments in promising local industries and services.[18] "Clusters"—the geographic concentration of interdependent, specialized firms working in similar or related activities—are a model drawn from the industrial development of Europe, the Americas, and Asia. Clusters can build on the strengths of local communities, including social capital and abundant labor, to overcome common constraints to economic expansion, such as weak financial markets and institutions.[19]

Because many of the existing clusters in the MENA region are labor intensive, they fit the comparative advantage of the rural populations. In many instances, these clusters develop organically without government intervention. For example, in an attempt to leverage existing rural strengths and generate productive employment opportunities for economic development, Egypt's government is working to promote the growth of the existing industrial clusters across the country, especially those in the poorest rural areas (for example, the El-Mahalla El-Kubra Textile Cluster, Robiky Leather Cluster, Damietta Furniture Cluster, and many other agro-processing clusters). All these clusters benefit from abundant cheap labor and densely populated communities.[20]

At the regional level, agriculture, food security, and natural resources are central to the recent "Beirut Declaration," adopted in June 2018 by the majority of Arab countries. The declaration aims to boost the supportive role of MENA governments in ensuring a key role for technology in achieving equitable and sustainable development.[21] Not surprisingly, energy, and in particular renewable energy (solar and wind mainly), is seen as one of the promising engines for sustainable and diversified growth in MENA's rural areas.[22]

Private sector employment rates in the MENA region are among the lowest in the world; the situation is worse in rural areas. Meanwhile, the scale of MENA countries' economic and fiscal vulnerabilities indicates that the public sector does not have the capacity to absorb the new labor market entrants.[23] Politicians are increasingly aware of the need to accelerate structural reforms to allow the private sector to flourish and generate the required jobs in rural areas. For example, Morocco's economic development policy targeting rural and lagging regions was updated in 2018 with a new territorial development approach that increases the voice of the regions in determining social and economic development policies—with a heavy reliance on public-private partnerships.[24]

With a shift toward rural revitalization, rural areas can play a key role in national economies by creating jobs, contributing to export earnings, and producing food and environmental benefits. For countries currently in conflict, the postconflict reconstruction era will provide an opportunity for the modernization of rural areas, especially in traditionally strong agricultural countries like Iraq, Syria, and Yemen. But revitalization will bear fruit only when the private sector begins to invest in rural areas. Budding initiatives to encourage entrepreneurship—with a special focus on agribusiness and on rural areas—that are ongoing in several countries, for example in Lebanon, may already be contributing to rural revitalization.[25]

OUTLOOK FOR 2019

In 2019, action at the national level—in peacebuilding, economic reform, and increased rural investment—is likely to be the key focus of MENA countries, as the lingering political divisions in the region will continue to stymie regionwide initiatives. Increased efforts in conflict countries should finally lead to peacebuilding and development interventions that will help stem the flow of displaced persons and contribute to peace. In nonconflict countries, governments should be deepening economic reforms and assessing the early results of recent reforms—especially in terms of targeting rural areas and increasing equitability. The renewed consensus around on- and off-farm rural revitalization should lead to more practical and localized implementation of policies and investments in emerging clusters, linking private and public sectors and taking advantage of the benefits of technological advances that allow for modernization and diversification.

CENTRAL ASIA

KAMILJON AKRAMOV, JARILKASIN ILYASOV, EVGENY TSVETNOV, AND ALLEN PARK

Kamiljon Akramov is a research fellow, and **Jarilkasin Ilyasov** and **Allen Park** are senior research analysts, Development Strategy and Governance Division, International Food Policy Research Institute, Washington, DC, USA. **Evgeny Tsvetnov** is a senior researcher at the Eurasian Center for Food Security, Moscow, Russian Federation.

Food security in Central Asia continues to be largely shaped by international commodity markets and developments in the region's major trading partners. Nevertheless, recent political events in the region had a positive impact on overall economic development as well as on food and nutrition security by stimulating intraregional trade and integration.

MAJOR DEVELOPMENTS IN 2018

Favorable external conditions led to better-than-expected economic growth in Central Asia in 2017. Increased demand for commodities boosted natural resource exporters while other economies in the region benefited from rising remittances and strong external demand.[1] However, in 2018, external economic conditions were mixed for Central Asian countries. Oil prices continued rising during the first three quarters of the year, but began to drop sharply in the fourth quarter. Wheat prices rose by about 15 to 18 percent during the year. Prices of the region's other commodity exports, such as metals, changed little in 2018.[2] In addition, geopolitical and trade tensions slowed economic growth in the region's key trading partners (Russia and China) and led to depreciation of regional currencies.[3] The Russian ruble depreciated by about 20 percent and the Kazakh tenge by nearly 13 percent against the US dollar during 2018. On balance, these external factors had a slight positive effect on economic activity and food security in Central Asia through their impact on export earnings, remittance flows, and investments from the region's main economic partners. Remittance flows from Russia to Kyrgyzstan, Tajikistan, and Uzbekistan increased by 3 to 7 percent in the first nine months of 2018 (valued in US dollars),[4] and depreciation of the national currencies of these countries meant even greater increases when converted to local currencies.

Depreciation of national currencies, however, also generated inflationary pressures in Central Asia's consumer markets, including food markets.[5] For example, in Uzbekistan, continuing price reforms and the lagged effects of official exchange rate devaluation triggered double-digit inflation in 2018. Ongoing efforts to reform utility and energy prices also led to increasing inflationary pressures in Turkmenistan. While no official data are available, reports of increasing food supply difficulties and shortages in Turkmenistan have appeared in the media since 2017.[6]

Recent political and economic developments in Central Asia led to a surge in intraregional trade and investment. A report from the Asian Development Bank finds that intraregional trade increased by 7.3 percent in 2017 over the previous year, and intraregional foreign direct investment increased by 5.5 percent.[7] Kazakhstan and Uzbekistan are the biggest trading partners within the region. Bilateral trade between these two countries increased by 31.2 percent to about US$2 billion in 2017, and by an additional 46.6 percent in the first three quarters of 2018. Likewise, movement of goods, capital, and people improved significantly between Tajikistan and Uzbekistan after the two countries ended visa requirements in March 2018, some 18 years after their introduction.[8] Direct bus and airline connections began operations between the two capital cities and other major cities in the two countries. And trade between the two countries increased more than twofold to US$283 million in the first three quarters of 2018, compared to the same period of 2017.[9]

China has become an important trading partner for the region. In 2017, Chinese trade with Kazakhstan reached US$1.3 billion; with Turkmenistan US$590 million; and with Kyrgyzstan US$567 million.[10] China has now replaced Russia as Uzbekistan's leading trade partner. In the first three quarters of 2018, trade between Uzbekistan and China reached US$4.4 billion, accounting for nearly one-fifth of Uzbekistan's total foreign trade.[11] China is also the largest buyer of Turkmenistan's hydrocarbon exports. However, Turkmenistan is trying to diminish the risk of overreliance on a single market by embarking on construction of the Turkmenistan-Afghanistan-Pakistan-India (TAPI) pipeline, considering the potential resumption of gas exports to Russia, and evaluating gas swaps with Iran.[12]

In recent years, Central Asian countries have begun exporting a wider variety of agricultural products to China. According to the World Bank, rising demand for food products in China and elsewhere in Asia may provide an opportunity for Central Asian countries to diversify their agricultural production and increase export earnings, especially from oilseeds, horticulture, and livestock.[13] Central Asian countries are working with Chinese trade authorities to promote agricultural exports: in September 2018, Chinese inspectors visited Kazakh grain and livestock producers to assess the feasibility of exports, and Uzbekistan began exporting cherries in 2018. Some observers suggest that Central Asian countries could take advantage of China's growing trade disputes with the United States to act as alternative suppliers to China, especially as transportation infrastructure develops and regional integration progresses.

MALNUTRITION

Multiple forms of malnutrition continue to burden the region. Although the prevalence of childhood stunting has declined by over 20 percentage points since the early 2000s in Tajikistan, about 17 percent of children under five are still stunted.[14] The latest available data suggest that nearly 12 percent of children under five in Turkmenistan are stunted.[15] Also prevalent among Central Asian populations are micronutrient deficiencies including anemia, which is estimated to affect 34 percent of women of reproductive age. At the same time, overweight and obesity are becoming more common. The number of overweight children under five in the region nearly doubled between 2000 and 2017, from 0.5 to 0.8 million.[16] In fact, the prevalence of overweight has overtaken wasting among children in all Central Asian countries except Tajikistan. While the obesity rate for Central Asian adults remains relatively low (16.4 percent in 2014), the prevalence of obesity grew by 41 percent between 2000 and 2014. About 7 million adults could be considered obese in 2014.[17]

To help Central Asian countries support a nutrition agenda for meeting the UN Sustainable Development Goals (SDGs), the countries of the region together with several UN agencies established the Central Asia and Caucasus Regional Nutrition Capacity Development and Partnership Platform. Central Asian governments also adopted laws on food security as part of their 2030 SDG agenda. In 2017, the government of Kyrgyzstan amended its law on food security to include the concept of a "healthy diet" in response to recent changes in its food security situation.

POLICIES RELATED TO RURAL REVITALIZATION

More than one-half of Central Asia's population resides in rural areas. Kazakhstan is relatively urbanized, with 57 percent of the population living in urban areas, but Kyrgyzstan and Tajikistan are predominately rural, with nearly 64 percent of Kyrgyzstan's population and 73 percent of Tajkistan's population living in rural areas. In both Turkmenistan and Uzbekistan, approximately half of the population lives in rural areas. Projections suggest that urbanization will continue in the region, but a considerable share of the population will continue to live in rural areas. Although agriculture's share in value added has declined considerably in recent years, it still accounts for a large share of employment due to significant sectoral productivity differences (Figures 1 and 2). The persistent productivity gaps between agriculture and nonagriculture sectors are reflected in the higher levels of poverty and malnutrition in rural areas in comparison with urban areas.[18] The majority of the poor in the region live in rural areas. For example, nearly 70 percent of the poor in Kyrgyzstan and more than 80 percent of the poor in Tajikistan live in rural areas.

Central Asian countries are promoting rural development and revitalization through rural infrastructure development, establishment of agriculture-based economic clusters, promotion of trade and service cooperatives, adoption of innovative technologies for agricultural production, and economic diversification in rural areas.

FIGURE 1 Agriculture's share in total employment

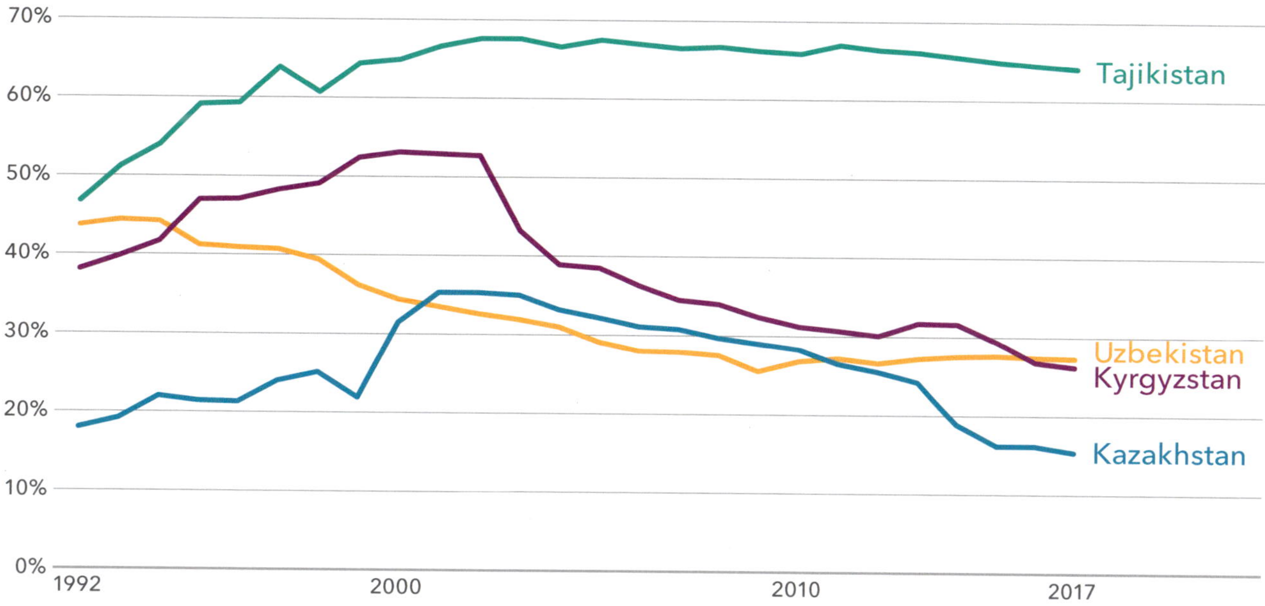

Source: Authors' computations using data from the World Bank's 2018 World Development Indicators database, the Asian Development Bank's Key Indicators for Asia and the Pacific 2018 database, and national statistical offices.

FIGURE 2 Agriculture's share in GDP

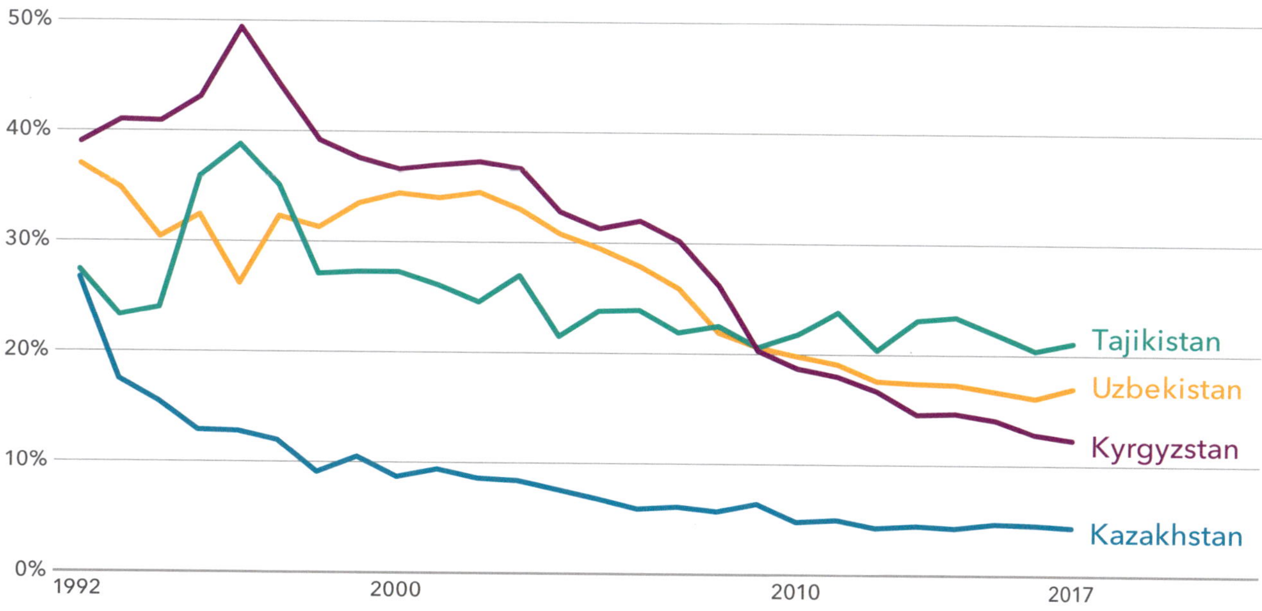

Source: Authors' computations using data from the World Bank's 2018 World Development Indicators database, the Asian Development Bank's Key Indicators for Asia and the Pacific 2018 database, and national statistical offices.

Efforts to develop specific regions or sectors of an economy through cluster development and special economic zones have had mixed results. The earliest efforts date to the 1990s, when the Kazakh government established nine free economic zones across the country to spur transition from its centrally planned economy to market-oriented development. Due to corruption, a lack of transparency, and an ineffective legal and regulatory framework, these economic free zones were replaced in 1996 by a new initiative known as special economic zones (SEZ).[19] Kazakhstan now has 10 SEZs, of which 6 have fully developed infrastructure. SEZs are intended to accelerate the development of competitive industries and the introduction of advanced technologies, attract investments, and create employment opportunities.[20]

Beginning in 2004, Kazakhstan adopted "cluster" development as a cornerstone of its industrial diversification strategy. However, a recent report suggests that although programs promoting SEZs and industrial zones (IZs) could potentially support the development of clusters, a disconnect exists between current SEZs and cluster development in Kazakhstan. Kazakhstan's SEZs and IZs are viewed as industrial infrastructure to attract investment, but their potential to create employment opportunities, enhance productivity and competitiveness, and promote cluster development is neglected.[21]

In Uzbekistan, development of the cluster approach in the food and agriculture sector is a newer phenomenon, more focused on both attracting foreign direct investment and making use of domestic industrial and production capacity. The country's policy of developing agriculture-based clusters, adopted in 2018, aims to promote processing and logistical centers for export of select agricultural products. A special presidential decree created clusters for production of horticultural and fruit products throughout the country, with specific focus on certain regions such as Bukhara and Andijan for grapes and wine products, and for fish products in various regions of the country.[22] Uzbekistan's cotton and textile industry is being restructured by expanding clusters for harvesting and processing cotton and producing textiles. The country plans to process all domestically produced cotton fiber and increase the share of finished textile products from its current 40 percent to 60 percent by 2020.[23] While the benefits of agriculture-based clusters—economies of scale, cost sharing, diffusion of innovations, and off-farm employment opportunities in rural areas—are well known, policymakers should be aware of potential limitations and risks, such as possible negative impacts on land tenure practices and smallholders' welfare.

Central Asian countries have been opening their economies and improving the business environment to increase competitiveness in regional and international markets. These efforts are reflected in the World Bank's annual *Doing Business* report, where all Central Asian countries (except Turkmenistan, which is not included in the ranking) improved their rankings in 2018 in terms of the ease of doing business, with the greatest progress reported for Kyrgyzstan. Kazakhstan ranked 28 among 190 economies in the list, thus becoming the first Central Asian country in the top 30. Nevertheless, considerable progress can still be made in other Central Asian countries, especially in Tajikistan, which remain at the lower end of the report's rankings.[24]

Agricultural diversification and development of competitive agricultural and food value chains are given priority in the region's national development strategies. Kazakhstan, for example, remains committed to its efforts to diversify production away from grain crops to other products, such as oil crops, by providing financial incentives to farmers and access to other forms of financial support through its joint-stock company KazAgroFinance.[25] Uzbekistan and Tajikistan are using both administrative and financial instruments to promote agricultural diversification.

Bilateral and multinational development partners, including the World Bank Group and the Asian Development Bank, remain committed to supporting these initiatives. Beyond their continuing support for infrastructure development, donor agencies have become more active in recent national economic diversification initiatives, including initiatives for agricultural diversification and for development of competitive agrifood value chains.

The Asian Development Bank plans to provide substantial financial support for investments in agriculture, natural resources, and rural development—pledging about US$620 million to support agricultural productivity and export orientation in

Kazakhstan for 2019–2021, about US$85 million to strengthen agricultural services and improve climate change adaptation and disaster resilience in Kyrgyzstan during the same period, and more than US$400 million to improve productivity of a diversified agriculture sector in Uzbekistan for 2018–2020.[26] The World Bank has pledged support for development of the region's horticultural sector and signed loan agreements for US$940 million with Uzbekistan in 2018 to finance four new projects focused on horticulture and on energy efficiency and emergency medical services.[27]

BUILDING ON REFORMS

Central Asian countries will continue to face global and regional risks in 2019 and beyond. At the global level, commodity price uncertainty and the pursuit of inward-looking policies in other regions may threaten the momentum of growth and amplify fiscal and external vulnerabilities. Regionally, reform fatigue or vested interests could threaten economic development efforts, particularly given the difficulty of persevering with structural reforms for long-term benefits. These uncertainties pose risks for economic growth as well as for food and nutrition security.

The working-age population is growing rapidly in the region, especially in Kyrgyzstan, Tajikistan, and Uzbekistan, but many of these people are either inactive, unemployed, or working abroad.[28] Given the uncertain environment and growing demand for employment, especially in rural areas, Central Asian countries should look to strengthen the role of the private sector in their economies by accelerating reforms and improving their institutional frameworks, and should continue to look for opportunities for regional integration.[29] This is especially important for agriculture and rural nonfarm development in countries that are still heavily rural, such as Kyrgyzstan, Tajikistan, and Uzbekistan. In agriculture, reforms to liberalize land tenure policies and create an enabling environment for collective management of common-pool resources are needed to promote the expansion of labor-intensive and high-value sectors such as horticulture. Such reforms should consider direct and indirect impacts of climate change on the availability of water resources for irrigation and the increasing frequency of extreme events such as droughts. These countries also should support women's participation in the workforce, given that women's labor participation rates remain markedly low in Central Asian countries.

In this regard, expected reductions in transaction costs may help to expand international and regional trade and boost growth and employment opportunities across the region. China's Belt and Road initiative is expected to significantly reduce shipment times and trade costs for Central Asian countries. If border delays are reduced by half, shipment times could fall by more than 25 percent along the China–Central Asia–West Asia Economic Corridor; likewise, trade costs could fall by over 21 percent along this economic corridor.[30]

These developments offer hope of significant socioeconomic progress in the region. If the governments in Central Asia can create an enabling environment for regional integration, trade, and private sector development, the payoff will be not only economic development but also a more competitive, healthy, and well-nourished population.

SOUTH ASIA

ANJANI KUMAR, ABDUL WAJID RANA, STEPHEN DAVIES, AKHTER AHMED, AND P. K. JOSHI

Anjani Kumar is a research fellow, South Asia Region, International Food Policy Research Institute (IFPRI), New Delhi, India. **Abdul Wajid Rana** is a program leader and **Stephen Davies** is a senior research fellow, Development Strategy and Governance Division, IFPRI, Islamabad, Pakistan. **Akhter Ahmed** is a senior research fellow and chief of party, Bangladesh Policy Research and Strategy Support Program, IFPRI, Dhaka, Bangladesh. **P. K. Joshi** is director for South Asia, IFPRI, New Delhi, India.

South Asia remained the fastest growing region in the world in 2018, as economic growth continued to strengthen. However, growth rates varied greatly across the region—exceeding 7.0 percent in Bangladesh, India, the Maldives, and Nepal, and reaching 5.8 percent in Pakistan, but only 2.6 percent in Afghanistan. Growth in most of the South Asian nations was driven primarily by domestic consumption, with limited contributions from exports and investments. Rising oil prices are a concern because the region is a net importer of oil; higher oil prices exert pressure on current accounts and add to fiscal deficits.

The structure of South Asia's national economies has changed dramatically in recent decades, marked by a sharp decline in the share of the agriculture sector. Agriculture now accounts for less than 16 percent of regional gross domestic product (GDP) in South Asia, varying from 5.9 percent in the Maldives to 27.0 percent in Nepal.[1] However, the share of the labor force in agriculture has declined more slowly, remaining substantially greater than agriculture's share in GDP—agriculture's share in employment across the region is about 43.0 percent, varying from 7.5 percent in the Maldives to 62.0 percent in Afghanistan. Although the share of agriculture in rural employment remains higher, growth of the rural nonfarm sector is accelerating. The rural nonfarm sector now provides a sizable share of rural income and employment, primarily in services, and has grown more quickly than agricultural employment—the rural nonfarm sector now generates about 60 percent of rural income in India and Nepal and 57 percent in Pakistan and Bangladesh.[2]

RURAL TRANSFORMATION

Despite an acceleration of urbanization in recent years, South Asia's demographic transition has been slow and the region remains predominately rural—67 percent of the population lives in rural areas.[3] In many countries, literacy rates remain low, especially in rural areas: about 60 percent of the rural population in South Asia is literate as compared to 79 percent of the urban population.[4] Although rural poverty has declined sharply, one-fifth of South Asia's rural population remains poor. The incidence of rural poverty is highest in Afghanistan (38.0 percent) and lowest in Sri Lanka (7.6 percent).[5] And while some countries, including Bangladesh and Nepal, have made impressive progress in reducing malnutrition, South Asia is still the biggest contributor to malnutrition in the world. More than one-third of children (about 60 million) under five years of age are stunted in South Asia, accounting for about 35 percent of the world's stunted children, and stunting rates are higher in rural areas than in cities.[6] Further, over 5 million children in South Asia are overweight, accounting for about 14 percent of the world's overweight and obese children.[7] South Asia is subject to multiple burdens of malnutrition—inadequate dietary energy and protein, micronutrient deficiencies, and, more recently, overweight and obesity, which have been increasing rapidly. These problems can coexist in the same household and even the same individual.[8] Although undernutrition remains more prevalent than obesity, governments, the private sector, and civil society would be well-advised to address this emerging problem now.[9]

MIGRATION AND REMITTANCES

Migration and remittances have direct and indirect effects on both the countries that migrants leave and their destination countries. For the countries of origin, the welfare benefits of migration are often sizable; remittances can reduce poverty and stimulate economic activity. In Nepal, a dramatic increase in remittances is credited with between one-third and one-half of overall poverty reduction; poverty fell from 42 percent in 1996 to 31 percent in 2004. Migration within and from South Asia has increased significantly over the past two decades, sparking employment, economic growth, and development in the destination as well as source areas. In 2018, South Asia registered an international migration of over 39 million people—17 percent of the world's total migration—and the region received US$132 billion in remittances. Remittances to the region have been increasing since 2000 and are the now the world's second highest after East Asia and the Pacific, making a substantial contribution to South Asian economies. A recent upsurge in remittances, which rose 13.5 percent in 2018, was driven by stronger economic conditions in the advanced economies, particularly the United States, and the positive impact of higher oil prices on remittances from Gulf Cooperation Council countries. For Nepal, remittances account for more than one-third of GDP; for Sri Lanka, remittances constitute an estimated 8.5 percent of GDP; and for Bangladesh and Pakistan, about 7.0 percent of GDP. Bangladesh and Pakistan registered a surge in remittance inflows of 18 and 6 percent, respectively. India receives the largest amount in international remittances (US$80 billion in 2018), accounting for 2.8 percent of GDP. Remittance flows to Afghanistan are officially about 1.8 percent of its GDP, but the actual flows are believed to be higher.[10]

BASIC SERVICES AND QUALITY OF LIFE IN RURAL AREAS

South Asian governments have increased their commitment to providing basic services in rural areas in recent years, and their efforts are showing results for rural residents (Figure 1). About 80 percent of the region's rural population now has access to electricity, a big increase from 57 percent in 2007. Rural areas in Bhutan, Pakistan, and Sri Lanka are almost

FIGURE 1 Access to basic services in rural South Asia

Source: World Bank, 2018 World Development Indicators, https://data.worldbank.org/.

entirely electrified. Access to electricity is lowest in Bangladesh (59 percent), followed by Afghanistan (77 percent) and India (78 percent). Basic sanitation services have also expanded, but as of 2016 only 36 percent of South Asia's rural population had access to these services (ranging from 32 percent in Afghanistan and India to more than 95 percent in the Maldives and Sri Lanka). About 53 percent of the population in rural South Asia had access to safe drinking water in 2016, up from 38 percent in 2002. The incidence of open defecation remains quite high at 47 percent, but is down substantially from 2002, when it stood at an alarming 70 percent.[11]

Recently, air pollution related to the burning of crop residues has generated concern in some countries (notably, India and Pakistan). This burning continues because it offers an easy solution to disposing of those residues, reduces pests and diseases, and allows for increased cropping intensity.[12] Although burning of crop residues is prohibited by law in many countries, enforcement is difficult. More pragmatic and coherent policies along with appropriate technological solutions, such as monitoring through satellite imagery, should be explored to minimize the burning of crop residues and reduce air pollution.

CHANGING CONSUMPTION PATTERNS AND CHALLENGES

Rice and wheat are the region's major staple crops, accounting for about two-thirds of total dietary energy. But food consumption patterns have changed in the region over the past few decades, and the changes are most apparent in rural areas—consumption of cereals is declining while consumption of animal-sourced foods, fruits and vegetables, and processed foods is increasing. Pressure to expand food production to meet growing demand is putting stress on natural resources. The resulting expansion and intensification of agriculture are leading to land degradation, deterioration of soil quality, and loss of biodiversity, potentially jeopardizing the region's capacity to meet future food demand. Agricultural growth also poses risks for water resources. Facing the world's lowest per capita renewable freshwater resources, millions of rural South Asians have benefited from the growing use of groundwater. But aquifers are being depleted and the water table is falling, particularly in India. Water quality is also deteriorating throughout the region due to nutrient overloads and industrial pollution, raising concerns about food safety and drinking water quality.

Nevertheless, the changing dynamics of consumption patterns—driven largely by urbanization, demographic transitions, increasing income, and growing integration of food supply chains and food systems—offer new opportunities for entrepreneurship and employment in rural areas. Similarly, climate change increases concerns about sustainability, but also offers opportunities for rural communities to generate benefits through improved management of natural resources.

POLICY DEVELOPMENTS

A rural transformation is taking place in South Asia because of changes in demography, urbanization, incomes, and livelihoods. Most countries in the region are reemphasizing and reorienting their rural development efforts toward revitalizing rural economies.

India, home to 70 percent of the region's rural population, has unveiled a number of measures to bolster the rural economy: The government will maintain minimum support prices for major crops (25 at present) equal to at least 1.5 times their production costs. The efficacy of this policy will depend on the extent of implementation, as government procurement has been largely restricted to rice, wheat, sugarcane, and cotton. Further, this policy risks creating barriers to private sector entry into agricultural markets. India will also upgrade 22,000 rural *haats* (local informal markets) to Gramin Agricultural Markets (GrAMs), and a fund of about $US350 million is proposed to develop and upgrade agrimarketing infrastructure. All Indian residences are expected to be connected by all-weather roads by early 2019 (Figure 2).

Other government measures to boost India's rural economy and improve rural livelihoods include development of cluster-based specialized farming, promotion of organic farming, support for farmers' organizations and processing through the Operation Greens scheme, extension of farmer credit to fisheries and animal husbandry farmers, assured purchase of solar energy generated by farmers, and establishment of funds for infrastructure for irrigation, dairy processing,

FIGURE 2 Road density in South Asia

Source: Land Portal, accessed December 10, 2018, https://landportal.org/book/indicator/fao-21017-6124.

Note: Road density for South Asia is calculated as the average of road density of South Asian countries, weighted by their respective geographical areas. Data for Afghanistan are for 2004 and 2010; Bangladesh, 1990 and 2003; Bhutan, 2001 and 2011; India, 1990 and 2011; Maldives, 2005; Nepal, 1999 and 2008; Pakistan, 1990 and 2011; Sri Lanka, 1990 and 2010.

aquaculture, and animal husbandry. Further, in 2018, India's government embarked on two ambitious schemes under the Healthy India initiative (Ayushman Bharat): the creation of 150,000 healthcare centers and the Prime Minister's Public Health Scheme (Pradhan Mantri Jan Arogya Yojana), which will provide health coverage to over 100 million poor and vulnerable families (approximately 500 million individual beneficiaries) of about US$8,500 for hospitalization per family per year. It will be the world's biggest health scheme.[13]

In Pakistan, despite political instability, several recent policies have created a framework for agricultural reform and rural revitalization. These include the 2018 National Food Security Policy and National Water Policy, the 2012 National Climate Policy and enabling legislation of 2017, the 2018 Provincial Agriculture and Water Policies, and the National Drinking Water Policy. The National Food Security Policy focuses on value-added growth in the agriculture sector for both domestic and export markets and raising overall rates of economic growth to promote food security. It sets a goal of 5 percent growth in the agriculture sector to reduce poverty and support a GDP growth rate of 7 to 8 percent. The policy also includes minimum standards for food safety, seed certification, and pest and animal health surveillance. Federally and provincially funded flagship programs—such as promotion of high-value crops, development of marginal and environmentally fragile areas, and targeted food distribution systems—have been initiated to create a modern, efficient, and diversified agriculture sector, with the aim of ensuring domestic food security and providing high-quality products for both domestic sale and export.

In response to its emerging water crisis and concerns about climate change, Pakistan recently announced a National Water Policy designed to enhance water storage capacity, improve water governance, increase efficiency of water use, and introduce climate-resilient agricultural practices, watershed management, better management of subsoil water, and hydropower development. Likewise, provincial agricultural policies focus on enhancing farm productivity, diversifying to high-value and climate-resilient agriculture, increasing value-addition and competitiveness, strengthening research and extension, phasing out wheat procurement and diverting resources to agriculture, and enhancing resilience of smallholder farmers to climate change and natural disasters.[14]

Bangladesh has emerged as a leader in improving rural development indicators as well as food and nutrition security. In 2018, Bangladesh earmarked 27 percent of its budget for development of social infrastructure. To maintain the momentum, Bangladesh initiated measures focused on health and nutrition in rural areas and on consolidation of agricultural growth. A health protection program was introduced in 2018 for people below the poverty line, and free physician consultancy services are now available 24 hours a day. Almost 14,000 healthcare providers have been recruited to staff 13,500 community clinics that bring health services directly to the rural poor and marginalized communities.

Bangladesh has long commited to providing affordable and quality health, nutrition, and family welfare services. Rural social service programs were first introduced in 1974 to reduce poverty. Today, interest-free microcredit of 5,000 to 30,000 taka (about US$65–375) per family is being distributed among the rural ultra-poor through a variety of social service and development programs; such microcredit schemes are generating self-employment opportunities. Bangladesh is also supporting the political, social, administrative, and economic empowerment of women by facilitating women's participation in the workplace—for example, by reducing violence against women and by providing access to information technologies to rural underprivileged woman. Bangladesh is also investing in rural roads, and recently signed a memorandum of understanding with the Asian Development Bank to upgrade about 1,700 kilometers of rural roads to all-weather standards. To support agriculture, Bangladesh imposes no import tariffs on primary inputs, including fertilizer, seeds, and insecticides.[15]

Other countries in the region are also pursuing rural revitalization. Nepal registered robust growth in agriculture and overall GDP in 2017. Following the country's transition from a centralized system of governance to a federal structure, the government began several initiatives to revitalize the rural sector, including programs for employment, women's empowerment, and agricultural modernization.[16] In Sri Lanka, major goals include development of rural economies and strengthening land ownership for the rural population. The Sri Lankan government is working to improve rural connectivity, enhance healthcare facilities, provide low-interest credit to farmers, launch a weather-based index insurance scheme, and create "fair price" shops in rural areas.[17]

CHALLENGES AND OPPORTUNITIES

The rural transformation in South Asia is at a crossroads. The rural sector has been gradually transforming, but agricultural livelihoods still provide more than 43 percent of employment, meaning that most jobs in South Asia are in the informal sector. Implementing a "decent employment agenda" will require improving rural livelihoods. The predominance of smallholder farming communities, their increasing political voice, and national commitments to the Sustainable Development Goals are prompting governments in the region to emphasize inclusive rural development strategies.

Looking forward to 2019, the prospects for rural development are encouraging. A general election is scheduled for early 2019 in India, with potential to increase attention to rural areas where the majority of voters live. Policy decisions in India have significant spillover effects on policy developments in other countries of the region. Across South Asia, inclusive employment growth, continued agricultural productivity growth, and strengthening of the agriculture-based rural nonfarm economy will be essential to ensure inclusive rural transformation. However, the region remains poorly integrated and is not taking advantage of its cultural affinities, common geography, or the advantages of proximity. In addition, escalating trade tensions and tighter global liquidity may constrain the prospects for South Asia in 2019.

EAST AND SOUTHEAST ASIA

KEVIN CHEN, PETER TIMMER, DAVID DAWE, AND ZIMEIYI WANG

Kevin Chen is a senior research fellow and head of the East and Central Asia Office, International Food Policy Research Institute (IFPRI), Beijing, China. **Peter Timmer** is professor emeritus, Harvard University, Cambridge, MA, USA, and a nonresident fellow, Center for Global Development, Washington, DC, USA. **David Dawe** is a senior economist and regional strategy and policy advisor, Food and Agriculture Organization of the United Nations, Bangkok, Thailand. **Zimeiyi Wang** is a research assistant, East and Central Asia Office, IFPRI, Beijing, China.

East Asian economies were buoyed by domestic consumption that stimulated growth in the first half of 2018, despite an external environment that was less favorable than in 2017.[1] The region is expected to remain on a positive trajectory through 2022, with economic growth in China and member states of the Association of Southeast Asian Nations (ASEAN) projected to average 6.2 percent and 5.2 percent, respectively.[2] This continued growth should lead to further declines in poverty. But a combination of factors—including escalating protectionism, China's slower growth, and adverse climate conditions in some countries—may prevent the region from meeting its full economic potential and further improving food security. In 2018, a year of both opportunities and challenges, countries worked toward deeper regional integration and rejuvenation of long-neglected rural areas to enhance preparedness and resilience amid growing uncertainty.

FOOD SECURITY AND NUTRITION TRENDS

East Asia has made substantial advances in food security. The number of undernourished people fell from 320.7 million in 2005 to 203.1 million in 2016. But in recent years, the rate of decline slowed significantly, and in 2017 the number of undernourished rose slightly to 203.3 million—124.5 million of whom live in China alone. The number of people suffering severe food insecurity, defined as those experiencing entire days without eating at times during the year, rose significantly, up from 48.9 million in 2015 to 82.2 million in 2017.[3] Considering the worsening global picture, these developments indicate that the region has considerable work to do to reach Zero Hunger.

Among factors that influence food security, climate variations in particular—especially increasingly frequent and intense weather-related disasters—are a growing concern. According to the Global Climate Risk Index, Myanmar, the Philippines, Viet Nam, and Thailand were among the countries most affected by extreme weather events in the past two decades.[4] In 2018, prolonged drought in Indonesia and an unusually heavy rainy season and related dam failures across Cambodia, Laos, Myanmar, the Philippines, and Viet Nam resulted in localized crop losses, particularly in agricultural heartlands where vulnerable subsistence farmers are concentrated.[5]

As countries strive to increase agricultural productivity and overall production, measures to mitigate the impacts of climate change on the rural sector are becoming institutionalized and often included in national plans.[6] Countries in East Asia are exploring adaptation and mitigation actions supported by research to develop climate-smart agriculture across the region, including increased education and capacity building for farmers, introduction of climate-resilient crops and cropping techniques, product diversification, and better access to meteorological information that can assist in weather forecasting and development of planting schedules. Risk-transfer mechanisms and social protection programs are also designed to enhance the resilience of communities vulnerable to climate change. The ASEAN Guidelines on Disaster Responsive Social Protection currently being developed are intended to help member states design and implement such programs. Cambodia, Indonesia, Malaysia, Myanmar, the Philippines, Singapore, Thailand, and Viet Nam have all either already established or are piloting agricultural insurance schemes.[7]

Rice remains a key staple in East Asia, although its dietary importance has decreased over time. Nevertheless, rice still dominates much of the region's agricultural landscape and helps to ensure livelihoods and freedom from hunger. Food security policies in the region therefore focus primarily on the staple, with some large importing countries such as China, Indonesia, Malaysia, and the Philippines aiming for food self-sufficiency and price stabilization, though at higher domestic price levels than in the region's rice exporters.[8] For low-income households, the current high prices caused by restrictive trade policies have clear negative impacts on food security. In the Philippines, domestic rice prices reached record levels after steady increases from the beginning of 2018. To help address the country's soaring inflation, the Philippine legislature passed a bill seeking to replace the quantitative restriction on rice imports with tariffs. How that law is implemented in 2019 will determine the impacts on rice farmers and consumers, food security, and agricultural transformation.[9]

Food insecurity—uncertain access to sufficient, safe, and nutritious food—can contribute to various forms of undernutrition, with particularly harmful impacts on the well-being of children. Despite impressive reductions, 19.7 million children in the region suffered from stunting and 6.7 million from wasting in 2017. In addition, overweight, obesity, and diet-related diseases are increasingly common as consumers in rapidly growing economies become more urban (and sedentary) and diets shift toward consumption of fats, meat, and sugar. Likewise, because poor, food-insecure families must prioritize cost when purchasing food, they often consume relatively cheap diets that are energy-dense but nutrient-poor.[10]

Tackling the challenges of food insecurity and malnutrition in all its forms is becoming a regional priority. Countries have implemented nutrition-sensitive interventions at national, district, and community levels, most of which identify and target the needs of vulnerable groups such as women and children. Some successful efforts include strengthening maternity protection in Viet Nam and Brunei, infant and young child feeding interventions in Cambodia, Laos, and Indonesia, and a weight management program for children in Singapore.[11] Progress has also been made at the regulatory and legislative levels, creating an enabling environment for the implementation of nutrition policies and programs. An emerging nutrition governance system in China is fostering political and administrative commitment to promoting agriculture-nutrition linkages for more nutritious and diversified crops. In 2017, ASEAN introduced its Regional Guidelines on Food Security and Nutrition Policy. The guidelines aim to ensure a sustainable and sufficient supply of affordable, safe, and nutritious foods by promoting the integration of policies for agriculture, trade, infrastructure, health, and education. The guidelines also advise member states to establish food-security and nutrition regulatory bodies and to adopt a more proactive stance to promote policy changes.[12]

UNCERTAINTY IN TRADE AND REGIONAL INTEGRATION

The US-China trade war is creating ripple effects worldwide, with agricultural markets for soybeans, pork, fruits, and nuts bearing the brunt of tariffs and retaliations.[13] Although the full impact of the turbulence has yet to be seen, the region already feels the heat.[14] If higher US tariffs weaken economic growth in China, the negative impact on growth throughout East Asia could be substantial.

The escalating tensions over trade are affecting soybean production and trade. China, which buys 60 percent of soybeans traded worldwide, is diversifying away from its usual trade partners—the United States, Brazil, and Argentina—and is boosting efforts to reduce imports by promoting soybean substitutes and increasing domestic production. Although subsidies to encourage soybean farmers have stimulated Chinese production, China is still projected to need imported soybeans to meet 87 percent of its domestic demand.[15]

Despite uncertainty, other economies in the region may find new opportunities in the turbulence. For example, exporting members of the Asia-Pacific Trade Agreement have benefited from China reducing tariffs on imports of 415 agricultural and related products to fulfill pledges to further open its consumer market. And Indonesia, Malaysia, Myanmar, Thailand, and Viet Nam may enjoy a boost in exports of some of their agricultural products, such as palm oil, cattle, fresh and processed fruits, and cashews, as China's retaliatory tariffs reduce imports of these products from the United States.[16]

Against the current backdrop of antiglobalization sentiment, most countries in the region recognize the long-term benefits of accessing a larger market and are working toward regional economic integration.[17] In 2017, the ASEAN Economic Community (AEC) adopted the AEC 2025 Consolidated Strategic Action Plan to foster regional integration. Major objectives include increasing trade and investment, integrating micro-, small-, and medium-sized enterprises into the digital economy, and developing an innovation-driven economy.[18] The Comprehensive and Progressive Agreement for Trans-Pacific Partnership has been ratified by 4 of 11 signatory countries and may take effect in early 2019. The ASEAN-led Regional Comprehensive Economic Partnership, which includes Australia, China, India, Japan, New Zealand, South Korea, and all 10 ASEAN members, took steps toward a year-end agreement on what could become the world's largest trading bloc.[19]

CREATING OPPORTUNITIES FOR RURAL AREAS

East Asia is undergoing the largest wave of urban growth in history, bringing huge social, economic, and environmental transformations. However, that progress has been uneven and insufficiently inclusive, with rural regions often left out of economic development. Recognizing this, governments in the region have pioneered various strategies to reverse rural decline in favor of a more resilient and sustainable economy, and some have made the issue of rural development central to their national goals.

A multidimensional rural revitalization strategy has been at the top of the Chinese government's agenda amid growing concerns over widening rural–urban disparities and the slowing economy. The strategy aims to modernize China's countryside and farm sector by 2035 by improving infrastructure, supporting technology innovations, integrating rural primary, secondary, and tertiary industries, improving rural public services, and deepening governance reforms in the countryside. It is estimated that China will increase funding for rural infrastructure development to about US$600 billion over the next four years. Key areas including transport, hydraulic engineering, energy, and information will benefit from joint efforts of relevant departments and different levels of government. The government is encouraging businesses to invest in the rapidly growing fields of e-commerce and information technology. Chinese Internet service providers, for example, have plans to invest US$2.16 billion in rural areas to improve 4G coverage and build fiber-optic networks.[20]

ASEAN countries are also implementing policies aimed at developing agriculture and rural areas. Myanmar launched its Agricultural Development Strategy to promote integrated value chain development, agricultural diversification, and institutional development. Cambodia has also put agriculture high on its national development agenda. The country's Agricultural Sector Strategic Development Plan 2014–2018 aims to keep the agriculture sector on its positive course of rising productivity, diversification, and commercialization while also promoting sustainable resource management and human development. Viet Nam will continue its national program to build "new-style" rural areas, which began in 2010. Under the program, more than 9,000 communes must meet 19 quantitative criteria set for the new rural areas, including criteria on planning, infrastructure, socioeconomic development, culture, and environment. The communes should assess their status against the criteria, support specific plans that identify priorities, and channel the necessary funds toward meeting their development needs. The program has already made progress in improving rural residents' well-being and increasing modernization.[21]

Several successful rural development efforts have called on the people themselves to take the initiative. For example, South Korea's New Village Movement (Saemaul Undong), launched in 1970, made the case that diligence, self-help, and cooperation, once catalyzed by government initiatives, could spark bottom-up efforts to improve the livelihoods of local communities, strengthen rural infrastructure, and protect farmers' rights. Similarly, Japan's One Village One Product program was introduced in 1979 to slow rural depopulation by encouraging rural residents to develop local products and industries with potential to be nationally, or even globally, marketable. By 2002, this program counted 338 local specialties among its products. Other countries in the region have established similar programs for rural development and poverty reduction—notably, Thailand's One Tambon One Product program, first established in 2001. By increasing employment opportunities and alternative income sources through community-based entrepreneurship, the Thai program has helped the rural workforce to grow and prosper. With the government's efforts to promote products of this program, sales reached 153 billion baht in 2017.[22]

LOOKING AHEAD

Looking forward, technological innovations in agriculture, especially the development of information and communication technologies (ICTs), will likely play a key role in transforming agricultural production and the lives of farmers. Countries including Thailand, Malaysia, and China are implementing national plans to encourage more efficient management and mechanization across the agricultural value chain.[23] A variety of e-services are expected to help the agriculture sector better leverage the power of the Internet. International organizations are also exploring the potential of ICTs to promote inclusive and sustainable food systems in the region.[24] The Asian Development Bank is supporting Cambodia's efforts to create a more effective environment for climate risk management and to reduce the agriculture sector's carbon footprint through use of ICTs. The Food and Agriculture Organization of the United Nations has been promoting an e-agriculture strategy in Laos, Myanmar, and the Philippines since 2015, and has worked with China to develop a Belt and Road Initiative umbrella program for innovation in food systems.

The sharing economy, online platforms, and other types of e-businesses have gradually won support in the region as effective tools to connect rural areas with broader markets. China allocated about US$654 million in 2017 to encourage e-businesses in rural areas with the goal of reaching 60 percent of impoverished counties. International e-retailers are beginning to make inroads in Southeast Asia, but the base remains small. If all goes according to plan, small- and medium-sized enterprises and farmers in the region will soon be able to sell fresh products directly to foreign customers, potentially contributing to poverty alleviation and rural rejuvenation.[25]

With growing urban markets, supportive domestic policies, and advances in technology, the revival of rural areas can facilitate agricultural transformation and help to make major economies more balanced by generating domestic demand. At the same time, ongoing trade disputes are threatening the regional economic outlook. Governments will therefore need to manage a complex array of policy tools to navigate uncertainty in an era of rising protectionism and to achieve Sustainable Development Goal 2, eradicating hunger by 2030.

LATIN AMERICA AND THE CARIBBEAN

EUGENIO DÍAZ-BONILLA AND VALERIA PIÑEIRO

Eugenio Díaz-Bonilla is head of the Latin America and Caribbean program, International Food Policy Research Institute (IFPRI), and **Valeria Piñeiro** is a senior research coordinator, Markets, Trade, and Institutions Division, IFPRI, Washington, DC, USA.

Global political and economic trends, including worrisome trade developments, as well as domestic change affected the economies and food and nutrition security of Latin America and the Caribbean (LAC) in 2018. According to recent estimates, the number of undernourished people in the region increased for the third consecutive year, reaching approximately 39 million, an increase of 400,000 since 2016. A primary cause was the deterioration in the food security situation in Venezuela. At the same time, the number of obese individuals in the region increased by more than 3 million annually over this period, and overweight and obesity have emerged as the main malnutrition and health problem.[1]

Repercussions are being felt around the region resulting from changes in US policies on trade and migration. Renegotiation of the North American Free Trade Agreement (NAFTA) led to the US-Mexico-Canada Agreement (USMCA) among the three partner countries, which has eliminated the threat of a trade war and larger disruptions in North American markets. The full implications of the agreement's new and expanded language on labor conditions, exchange rates, tariff rate quotas, intellectual property rights, and arbitration are still being analyzed, but the changes are likely to shape the US position in other trade agreements. The evolving "trade war" between the United States and China is also expected to affect the region. Analysis of the potential impact of this dispute, under several possible scenarios, finds that US–China trade disputes will reduce overall welfare for both countries. But other countries, including several in LAC, may benefit, at least initially, from trade diversion that increases demand for their exports.[2] Nonetheless, the region, and in fact the whole world, will suffer if those disputes escalate and depress global growth through trade and financial channels.

US migration policies are also affecting many countries in the region, particularly those in Central America. Large-scale deportations from the United States will hurt growth in these countries, increase poverty, and generate fiscal imbalances. But deportations also have a negative impact on the US economy.[3] Humanitarian crises such as the recent events related to a caravan of migrants, primarily from Honduras, that moved through Mexico toward the United States, will most likely continue to generate political tensions for all countries involved.

Recent elections in Mexico and Brazil are changing the political landscape in the region's two largest countries. At the end of 2018, the incoming government in Mexico announced its intention to invest in revitalizing rural areas, emphasizing the need to support small and family farms, particularly in the poorest regions in the south of the country. If those same efforts can be extended through regional cooperation to help Central American countries, as proposed by the new Mexican government, this could reduce migratory flows that pass through Mexico's territory to the United States. In Brazil, the new authorities have announced policies that may significantly reduce environmental safeguards in agricultural production, including proposals to leave the Paris Climate Agreement and to merge the Ministry of Environment with the Ministry of Agriculture. As of early 2019, it is not clear what the final position will be. Brazil's president-elect also suggested that the international community should compensate Brazil for continuing to protect lands now designated as reserves. Furthermore, economic advisors from the elected government indicated that they intend to take a broader approach to trade agreements and downplay the role of the regional trade bloc, MERCOSUR.

In Argentina, economic conditions deteriorated, forcing a strong devaluation of the peso, which lost about half its value between January and December 2018. While this adjustment should have helped to boost agricultural exports, its impact was offset in part by an increase in export taxes on a variety of products. Nevertheless, the net result has been an improvement in international competitiveness for Argentina's agrifood exports.

The humanitarian disaster in Venezuela continues to deepen. Five years of recession have reduced gross domestic product (GDP), which for 2018 is estimated to be roughly half of its 2013 value. Annual inflation may reach 1,370,000 percent by the end of 2018 (comparable to the worst historical hyperinflation, such as Germany in the 1920s and Zimbabwe more recently).[4] On average, about 60 percent of Venezuelans lost 11 kilograms of body weight in 2017, after losing 8 kilograms in 2016.[5] More than 40 percent of the population depends on government-provided food boxes. In addition to excluding a large part of the population, this food assistance program has allegedly been used to coerce political support in elections; accusations have also been made of corruption in the government's food procurement for the program.

In Nicaragua, social and political protests show no likelihood of resolution yet. Over the course of 2018, changes of government occurred in Colombia, Chile, and Honduras. In the first two countries, the new authorities announced departures from previous policies, which in the case of Colombia relate to peace agreements with guerrilla groups.

In regional developments, Argentina held the presidency of the G20 in 2018. As reflected in the theme for the year, "Building Consensus for Fair and Sustainable Development," Argentina's presidency aimed to help generate equal opportunities for all people and advance efforts to eradicate poverty, focusing on infrastructure, development, and a sustainable food future. The Ministers of Agriculture of the G20 issued a joint communiqué in May highlighting international cooperation on a series of topics, starting with soil management (the special topic selected by the Argentine government for its presidency) and also including trade, climate change, antimicrobial resistance, and reduction of food waste and loss.

LATIN AMERICAN AGRICULTURE AND RURAL REVITALIZATION

LAC has a long history of rural development approaches with a territorial focus, including the agrarian reform and settlement schemes of the 1950s and 1960s and "integrated rural development" in the 1970s and 1980s.[6] For much of these decades, rural development strategy subordinated agriculture to urban needs and to industrialization. Policy changes in the 1990s may have changed relative macroeconomic incentives in favor of the agriculture sector, but also resulted in the downplaying of specific policies and territorial strategies for agriculture and rural development. More recently, the concept of a territorial and multisectoral approach to rural development has been revived, for example in the 1999 Inter-American Development Bank Strategy for Agricultural Development.[7]

In general, territorial approaches, old and new, aim to stimulate the demand for products and services in local markets by strengthening urban–rural linkages. This entails promoting investments in infrastructure to establish linkages to local markets and foster intersectoral linkages and helping to develop small- and medium-sized agricultural and nonagricultural rural enterprises, including services.[8] In LAC, this approach has been accompanied by a process of decentralization and devolution of activities to local governments.

Urbanization plays a key role in shaping rural development in LAC. Among developing regions, LAC is the most urbanized: the urban population share is now almost 79 percent, well above the world average of 54 percent. While there are several megacities, almost 60 percent of the region's urban population lives in smaller urban agglomerations of fewer than one million inhabitants.[9]

The process of urbanization, along with the expansion of infrastructure and growing integration of rural and urban markets, has led to important changes in rural labor markets and nonagricultural rural activities, as well as the expansion and increasing complexity of agrifood value chains, both for domestic markets and for exports. Notably, small- and intermediate-sized towns have developed the strongest links to rural spaces and have had the greatest impact on agricultural value chains and the structure of food production, employment, and consumption. The growth of cities means that urban food security must be considered along with traditional concerns about food security in rural areas.

Rural areas in LAC enjoy access to basic services—electricity, water, and sanitation—that is above the average for developing countries and the world (Figure 1). In addition, the gap between rural and urban service coverage in the LAC region (measured as the ratio between rural to urban coverage) is smaller than for developing countries and the world as a whole (with the exception of drinking water), reflecting national and local investments and the pattern of urban development.

During recent decades, value chains also evolved significantly, with different waves of changes in the processing and retail segments reflecting increased urbanization and the large presence of intermediate cities. Agricultural and agro-industrial activities are increasingly controlled by large agricultural operators, input companies, agro-industrial processors, and supermarket chains. Farmers have become more involved in markets and commercialization.

In the processing sector, in addition to the traditional milling and meatpacking industries, a more recent expansion of large conglomerates has occurred, particularly in beef, poultry, and pork production.

FIGURE 1 Access to basic services

Source: FAOSTAT, accessed October 30, 2018, http://www.fao.org/faostat/en/#data.

The larger South American countries (Brazil, Argentina)—which were more advanced in terms of development and urbanization than their smaller Latin American peers—saw a surge of foreign direct investment in the processing sector in the mid-1980s and early 1990s (following liberalization and privatization in those years). Since then, the expansion of these investments related to processed and packaged products has transformed several value chains.[10] On the input side, international seed companies and other providers have expanded in the region, providing technology mainly for cereals and oilseeds. Machinery and irrigation companies have also extended their operations.

On the retail side, supermarkets have driven a restructuring of entire food chains, including processed and fresh products.[11] LAC is the developing region of the world with the earliest and widest expansion of supermarkets.[12] Retail transformation started in the early 1990s in Brazil and Argentina; Mexico and other countries followed later, in the mid-to-late 1990s.[13]

However, much is still needed to ensure dynamic urban-rural linkages and to ensure that revitalization of the rural sector reaches everyone, including vulnerable ethnic groups (Native Americans and Afro-descendants) and regions (such as the Dry Corridor in Central America, the Brazilian Northeast, and sections of the Andes) that are not receiving adequate support.

PROSPECTS FOR 2019

Regional developments in 2018 were framed by deteriorating world conditions related to the increases in interest rates in the United States, a softening of the world economy, escalating trade disputes, the reversal of capital flows from developing countries and emerging economies, and geopolitical uncertainties. Those negative trends are likely to deepen in 2019, a year that many observers expect to be difficult not only for the region but for the world as a whole. In view of a potentially deteriorating global scenario, LAC countries should tailor their macroeconomic, trade, and agricultural policies to strengthen their social resilience in a complex and uncertain environment.

"Now the time has come for *rural revitalization*—to make rural areas more productive, sustainable, healthy, and attractive places to live."

FOOD POLICY INDICATORS: TRACKING CHANGE

DECISIONMAKERS AND POLICY ANALYSTS NEED SOLID EVIDENCE and timely information to develop and implement effective food policies. The International Food Policy Research Institute (IFPRI) develops and shares global public goods—including datasets, indicators, and indexes—as part of its mission to provide research-based policy solutions that sustainably reduce poverty and end hunger and malnutrition. This information can be used to gauge the impact of policy changes and the progress made on specific aspects of development.

This section provides updates on data generated by IFPRI research. Indicators include investments in agricultural research, public spending on agriculture, capacity for food policy research, agricultural total factor productivity, and projections for agricultural production, food consumption, and risk of hunger to 2030 and 2050. All indicators are available online with an interactive display of the data.

Agricultural Science and Technology Indicators (ASTI)

Policymakers recognize that increased investment in agricultural research is key to increasing agricultural productivity. Despite this, many low- and middle-income countries struggle with capacity and funding constraints in their agricultural research systems.

Agricultural Science and Technology Indicators (ASTI), facilitated by the International Food Policy Research Institute (IFPRI) within the portfolio of the CGIAR Research Program on Policies, Institutions, and Markets, works with national, regional, and international partners to collect time series data on the funding, human resource capacity, and outputs of agricultural research in low- and middle-income countries. Based on this information, ASTI produces analysis, capacity-building tools, and outreach products to help facilitate policies for effective and efficient agricultural research.

TRENDS IN CAPACITY AND INVESTMENT

Global investment in agricultural research, once heavily weighted toward the developed world, has shifted dramatically in recent years toward the developing world. Whereas spending growth in high-income countries as a group has stalled since the turn of the millennium, the developing world has accelerated its agricultural research investments at a rapid pace, driven by high growth rates in China and India.

Agricultural research spending and capacity in Latin America and the Caribbean and in Asia have grown rapidly since 2000, but considerable differences remain across countries. Brazil's world-class research infrastructure and outputs contrast sharply with the lagging infrastructure, investment levels, and capacity in many Central American and Caribbean island nations. China accounts for most of the agricultural research spending growth in Asia, with India close behind. But underinvestment in countries such as Cambodia, Lao PDR, and Pakistan impedes their ability to respond to the threats to food security associated with widespread poverty, rapid population growth, climate change, and environmental degradation.

Although agricultural research spending and human resource capacity in Africa south of the Sahara have grown considerably, this growth has been uneven and trends are driven by large countries such as Ethiopia, Nigeria, and South Africa. Furthermore, many countries are overly dependent on volatile and unsustainable donor funds. The region is dealing with serious challenges on the human capacity side: long-term recruitment restrictions have left many research agencies with aging pools of researchers.

Female scientists remain severely underrepresented in agricultural research, despite their being in a unique position to address the pressing challenges of African farmers, the majority of whom are women. ASTI's Women in African Agricultural Research Data Portal—developed in partnership with African Women for Agricultural Research and Development (AWARD) and released in 2017—features detailed sex-disaggregated data on human resources in agricultural research, as well as graphic tools for country comparisons and analysis that allow researchers and policymakers to set priorities, establish benchmarks, and monitor progress over time. These data are currently for African countries, with plans to expand to other ASTI countries in the future. Such information is needed to understand the underlying obstacles facing female agricultural researchers, and how to overcome them.

The West Asia and North Africa region has made valuable progress in agricultural research investment since the 2008 global food crisis, but inadequate systems, funding, and human resource capacity—coupled with a lack of political stability—hamper food security. Many national agricultural research institutes need to improve pay, working conditions, and incentives to compete with universities and attract, retain, and motivate well-qualified researchers.

In all regions, the imminent retirement of highly experienced agricultural researchers without adequate plans for their replacement creates concern about the quality of future research outputs.

INDICATORS

"Agricultural research" includes government, higher education, and nonprofit agencies, but excludes the private for-profit sector. Total agricultural research spending includes salaries, operating and program costs, and capital investments for all agencies, excluding the private for-profit sector, involved in agricultural research in a country. Expenditures are adjusted for inflation and expressed in 2011 prices. Purchasing power parities (PPPs) measure the relative purchasing power of currencies across countries by eliminating national differences in pricing levels for a wide range of goods. PPPs are relatively stable over time, whereas exchange rates fluctuate considerably. In addition to looking at absolute levels of agricultural research investment and capacity, another way of comparing commitment to agricultural research is to measure research intensity—that is, total agricultural research spending as a percentage of agricultural output (AgGDP).

"Total agricultural researchers" (excluding the private for-profit sector) are reported in full-time equivalents (FTEs) to account for the proportion of time researchers actually spend on research activities. A critical mass of qualified agricultural researchers is crucial for implementing a viable research agenda, for effectively communicating with stakeholders, and for securing external funding. Therefore, it is important to look at the share of PhD-qualified researchers. Gender balance in agricultural research is important, given that women researchers offer different insights and perspectives that can help research agencies more effectively address the unique and pressing challenges of female farmers. Age imbalances among research staff should be minimized to ensure the continuity of future research as researchers retire.

MORE INFORMATION

Only a fraction of the available ASTI indicators are presented here. The ASTI website offers additional indicators, including national-level time series data on researcher capacity by qualification level, age bracket, discipline mix, and commodity, as well as a detailed breakdown of agricultural research investment. Interactive pages on the ASTI website allow users to access country-level time series data, make cross-country comparisons, create graphs, and download country datasets. The country pages also feature recent ASTI factsheets, other country-level publications, and detailed institutional information on agencies involved in agricultural research. ASTI's regional interactive benchmarking tools are convenient map-based instruments that allow users to make cross-country comparisons and rankings based on a wide set of financial and human resource indicators. The detailed ASTI datasets are available in an easy-to-use data download tool. Detailed information on definitions, methodology, and calculation procedures are available on ASTI's website.

VISIT ONLINE
www.asti.cgiar.org

CONTACT
ASTI (asti@cgiar.org)

TABLE 1 Agricultural science and technology indicators

Countries by region	Year	Agricultural research spending: 2011 PPP dollars (million)	Agricultural research spending: 2011 US dollars (million)	Agricultural research spending: As share of AgGDP (%)	Agricultural researchers (FTEs): Total	Agricultural researchers (FTEs): Govern-ment (%)	Agricultural researchers (FTEs): Female (%)	Agricultural researchers (FTEs): PhD-qualified (%)	Agricultural researchers (FTEs): Older than 50 (%)
AFRICA SOUTH OF THE SAHARA									
Benin	2016	30.5	13.8	0.60	201.5	39.2	16.8	62.1	32.7
Botswana	2016	17.5	9.6	2.27	116.0	59.0	29.1	30.3	23.5
Burkina Faso	2014	48.5	21.9	1.01	310.8	88.6	19.3[a]	52.5	39.1
Burundi	2016	10.9	3.7	0.39	134.0	63.4	15.2	18.7	23.8
Cabo Verde	2016	3.3	2.0	1.17	24.7	89.1	37.2	13.4	72.9
Cameroon	2014	45.9	22.1	0.34	240.1	80.7	20.9	40.8	34.0
Central African Rep.	2016	4.9	2.6	0.40	123.1	64.7	16.0	9.7	25.7
Chad	2016	5.9	3.2	0.05	89.1	79.3	4.3	18.4	22.4
Congo, Dem. Rep.	2016	27.7	15.7	0.24	553.0	64.3	10.7	16.4	26.6
Congo, Rep.	2016	6.3	3.9	0.26	79.2	63.8	18.3	37.9	35.4
Côte d'Ivoire	2014	82.1	39.7	0.53	253.2	62.1	16.9	71.3	36.3
Eritrea	2011	2.9	1.1	0.30	40.8	5.9	7.1	28.4	22.5
eSwatini	2016	6.7	3.6	0.70	26.2	39.7	27.9	38.2	34.7
Ethiopia	2016	162.1	47.2	0.29	3,024.6	88.4	10.5	7.8	6.7
Gabon	2016	1.6	1.1	0.10	65.0	73.7	21.4	25.2	29.4
Gambia	2016	4.8	1.6	0.88	58.8	84.4	13.4	11.5	21.2
Ghana	2016	178.6	82.6	0.91	598.9	66.9	21.3	44.8	36.2
Guinea	2016	3.9	1.5	0.17	262.4	85.2	8.1	14.7	57.8
Guinea-Bissau	2011	0.2	0.1	0.02	9.0	100.0	–	–	66.7
Kenya	2016	222.4	85.9	0.48	1,156.2	69.5	29.8	39.6	45.8
Lesotho	2016	2.8	1.5	0.94	33.2	81.3	54.2	13.9	15.1
Liberia	2011	6.7	3.5	0.51	45.1	71.0	20.4	10.6	24.6
Madagascar	2016	10.4	3.4	0.14	214.3	72.4	36.7	44.9	56.2
Malawi	2014	28.1	13.7	0.53	158.3	55.5	20.5	32.8	na
Mali	2016	57.8	25.7	0.44	295.6	85.2	14.5	39.1	60.5
Mauritania	2016	18.5	7.6	0.49	102.0	85.7	17.4	20.0	3.5
Mauritius	2016	37.3	20.7	4.82	141.8	70.7	42.0	15.2	43.9
Mozambique	2016	31.8	17.5	0.43	386.1	76.1	37.0	13.0	18.5
Namibia	2014	38.8	24.9	3.09	99.7	70.0	44.0	13.8	22.2
Niger	2016	22.2	10.4	0.32	199.9	77.5	14.9	39.1	33.6
Nigeria	2014	433.5	209.5	0.22	2,975.5	na	29.4[b]	23.7[b]	21.7[b]
Rwanda	2016	27.3	11.8	0.44	148.9	63.1	23.7	24.1	14.6
Senegal	2016	51.5	25.8	0.89	144.7	71.5	22.7[a]	76.5[a]	38.2[a]
Sierra Leone	2014	15.3	5.5	0.24	123.7	88.1	20.2[b]	13.7	24.8[b]
South Africa	2014	417.4	274.4	2.78	811.3	76.5	na	na	na
Tanzania, United Rep.	2016	68.5	23.0	0.17	785.0	69.8	29.4	26.9	32.0
Togo	2016	8.7	3.9	0.20	109.6	69.8	6.8	35.5	36.8
Uganda	2016	99.4	32.8	0.62	558.7	60.4	29.9	35.5	19.6
Zambia	2014	26.9	13.2	0.51	245.6	76.7	28.5[a]	14.1	18.1
Zimbabwe	2014	43.4	21.9	1.44	208.7	57.5	31.0	17.6	15.8

Table 1 continued

Countries by region	Year	Agricultural research spending: 2011 PPP dollars (million)	2011 US dollars (million)	As share of AgGDP (%)	Agricultural researchers (FTEs): Total	Government (%)	Female (%)	PhD-qualified (%)	Older than 50 (%)
ASIA									
Bangladesh	2012	250.6	78.2	0.37	2,121.0	78.0	12.4	35.3	28.8
Cambodia	2010	22.4	7.4	0.18	284.4	55.1	21.9	5.9	4.0
China	2013	9,366.2	5,081.5	0.62	na	na	na	na	na
India	2014	3,298.4	1,067.8	0.30	12,746.6	38.7	18.3	73.2	32.8
Lao PDR	2010	na	na	na	227.2	84.1	na	6.5	21.8
Malaysia	2010	592.3	282.5	0.99	1,609.4	84.1	49.2	24.9	29.3
Nepal	2012	53.4	17.8	0.28	403.4	85.8	12.5	14.8	43.0
Pakistan	2012	332.5	93.7	0.18	3,678.3	85.5	10.8	20.7	27.6
Sri Lanka	2009	61.8	21.6	0.34	618.8	88.5	46.9	24.2	na
Viet Nam	2010	136.0	44.5	0.18	3,744.2	74.4	na	17.8	na
LATIN AMERICA AND THE CARIBBEAN									
Antigua and Barbuda	2012	1.0	0.7	2.98	7.5	93.3	27.2	31.6	47.4
Argentina	2013	732.1	474.7	1.29	5,824.5	50.2	45.2	20.8	38.7
Barbados	2012	1.3	1.3	2.01	9.9	70.7	na	na	na
Belize	2012	2.3	1.3	0.66	12.6	35.7	23.0	1.6	23.3
Bolivia	2013	58.9	25.0	0.93	190.3	28.0	17.7	11.0	33.3
Brazil	2013	2,704.0	2,377.9	1.82	5,869.4	71.3	37.1	72.5	44.4
Chile	2013	186.4	134.1	1.65	715.7	66.7	33.3	36.8	29.2
Colombia	2013	253.7	159.5	0.79	1,102.9	40.3	36.1	22.5	25.5
Costa Rica	2012	37.1	25.5	1.06	241.5	25.9	34.3	14.0	53.1
Dominica	2012	0.2	0.1	0.18	3.0	–	33.3	33.3	66.7
Dominican Rep.	2012	20.3	10.3	0.30	199.6	75.5	24.2	10.3	44.5
Ecuador	2013	27.3	14.4	0.18	149.4	72.6	17.3[a]	9.6	23.4[a]
El Salvador	2006	6.6	0.4	0.15	76.9	78.0	na	na	na
Grenada	2012	0.4	0.3	0.71	1.8	83.3	na	na	na
Guatemala	2012	15.6	7.3	0.14	141.8	54.3	20.0	9.6	41.5
Honduras	2012	7.5	3.9	0.17	87.6	24.3	13.6	5.7	34.1
Jamaica	2012	11.8	7.4	0.89	62.1	66.2	47.2	21.6	na
Mexico	2013	710.4	438.8	1.05	3,967.4	38.7	25.3	47.5	54.8
Nicaragua	2012	17.5	7.0	0.38	131.5	60.9	29.7	8.8	35.8
Panama	2012	15.5	8.5	0.74	133.0	87.8	17.7	7.5	56.1
Paraguay	2013	26.8	14.2	0.26	209.5	57.8	37.2	5.4	28.1[b]
Peru	2013	83.4	46.1	0.35	339.1	34.3	27.2	13.1	52.9
St. Kitts and Nevis	2012	0.8	0.5	5.13	4.5	2.2	82.2	4.4	41.1
St. Lucia	2012	0.3	0.2	0.63	2.2	–	9.1	–	50.0
St. Vincent and the Gren.	2012	0.7	0.5	1.07	2.5	80.0	na	na	na
Trinidad and Tobago	2012	18.0	11.0	7.82	83.0	58.1	43.3	22.7	25.7[b]
Uruguay	2013	77.4	61.3	1.40	371.9	50.9	40.5	26.1	36.3
Venezuela	2013	86.2	54.5	0.31	503.1	83.1	48.2	16.3	39.9

Table 1 continued

Countries by region	Year	Agricultural research spending			Agricultural researchers (FTEs)				
		2011 PPP dollars (million)	2011 US dollars (million)	As share of AgGDP (%)	Total	Government (%)	Female (%)	PhD-qualified (%)	Older than 50 (%)
WEST ASIA AND NORTH AFRICA									
Algeria	2012	91.6	38.3	0.21	593.4	69.8	51.3	23.0	37.5[b]
Egypt	2012	528.4	144.7	0.44	8,419.7	86.1	36.3	67.6	na
Jordan	2012	36.2	15.0	1.84	272.3	72.5	18.3	35.5	39.7
Lebanon	2012	38.2	21.3	0.95	209.2	58.4	48.2	44.6	13.0
Morocco	2012	147.3	442.3	0.49	556.3	72.4	23.3	40.0	39.0
Oman	2012	110.0	2.6	6.51	243.6	91.5	31.1	25.5	10.1
Sudan	2012	57.3	26.3	0.14	932.8	72.9	na	36.9	20.3[b]
Tunisia	2012	63.0	97.1	0.64	541.6	80.0	32.7	61.8	38.3
Turkey	2012	537.3	376.7	0.51	3,009.4	60.4	32.5	41.6	11.1
Yemen	2012	38.7	13.7	0.56	526.7	78.7	7.1	28.7	27.0

Notes: na = not available. PPP = purchasing power parity. AgGDP = agricultural gross domestic product. (–) = zero. FTE = full-time equivalent.

Table only includes countries where ASTI has conducted survey rounds since 2002 (with the exception of China, for which data is based on official Science and Technology statistics). Agricultural research includes government, higher education, and nonprofit agencies, but excludes the private for-profit sector.

a = data exclude the nonprofit sector. b = data exclude the higher education sector.

Statistics on Public Expenditures for Economic Development (SPEED)

The Statistics on Public Expenditures for Economic Development (SPEED) database is a resource of the International Food Policy Research Institute (IFPRI) that contains information on agricultural and other sectoral public expenditures in 147 countries from 1980 to 2016.

Policymakers, researchers, and other stakeholders can use this robust database to examine both historical trends and the allocation of government resources across sectors. It also allows for comparisons among countries within a region or at a similar level of development. Because the SPEED database covers many countries for a long time period, it allows analysts of government spending to examine national policy priorities, as reflected in the allocation of public expenditures, and track development goals and the cost-effectiveness of public spending both within and across countries. In addition to the dataset in the form of spreadsheets, SPEED includes user-friendly tools that enable stakeholders to generate charts and geographic expenditure maps using menu-driven options.

Indicators reported include total agricultural expenditure, agricultural spending per capita, and the ratio of agricultural spending to agricultural gross domestic product (GDP) for the period 1980 to 2016. IFPRI researchers have compiled data from multiple sources, including the International Monetary Fund, World Bank, United Nations, and national governments, and conducted extensive data checks and adjustments to ensure consistent spending measurements over time that are free of exchange-rate fluctuations and currency denomination changes.

Agricultural spending in recent years has been highest in East Asia and the Pacific, largely driven by rapid growth in agricultural spending in China since 2005, as reflected by three key indicators of government expenditures: per capita agricultural spending, agricultural spending as a share of agricultural GDP, and agricultural spending as a share of total spending (Table 2). The Middle East and North Africa as well as Europe and Central Asia regions have spent large amounts on agriculture both per capita and as a share of agricultural GDP, but not as a share of total spending. Among developing countries, agricultural spending per capita and as a share of agricultural GDP in 2016 was lowest in Africa south of the Sahara; as a share of total spending, agricultural spending was lowest in Latin America and the Caribbean.

DOWNLOAD DATA
https://doi.org/10.7910/DVN/F7IOM2

CONTACT
Ifpri-speed@cgiar.org

TABLE 2 Agricultural public expenditure for economic development, by country

Region/country	Agricultural expenditure (billions 2011 constant US dollars)			Agricultural expenditure (billions 2011 PPP dollars)			Per capita agricultural expenditure (2011 constant US dollars)			Per capita agricultural expenditure (2011 PPP dollars)			Ratio of agricultural expenditure to agricultural GDP (%)			Share of agriculture in total expenditure (%)		
	1995	2005	2016	1995	2005	2016	1995	2005	2016	1995	2005	2016	1995	2005	2016	1995	2005	2016
EAST ASIA AND THE PACIFIC	**30.78**	**63.55**	**280.49**	**65.43**	**123.60**	**523.02**	**18.30**	**34.26**	**170.66**	**38.91**	**66.64**	**318.22**	**6.44**	**9.20**	**26.02**	**8.11**	**6.08**	**9.28**
China	15.71	52.92	269.25	28.96	97.54	496.28	13.04	40.59	195.30	24.03	74.82	359.97	4.79	11.24	29.18	8.43	7.22	9.90
Fiji[c]	0.03	0.03	0.01	0.06	0.05	0.02	42.32	38.49	15.24	72.81	66.21	26.22	6.74	7.45	3.31	4.01	3.26	1.24
Indonesia[a]	6.74	2.83	5.32	16.39	6.89	12.94	34.22	12.50	20.61	83.21	30.41	50.12	8.07	3.40	3.57	9.43	2.53	2.71
Malaysia	1.60	1.88	1.33	3.36	3.95	2.79	78.09	73.37	42.65	163.75	153.86	89.44	8.66	10.02	4.04	5.10	3.83	2.04
Mongolia	0.02	0.03	0.05	0.04	0.07	0.12	7.87	12.26	16.55	18.53	28.89	38.99	1.34	2.41	2.93	2.12	1.90	0.90
Myanmar	0.28	0.36	1.40	0.64	0.84	3.24	6.39	7.46	26.45	14.82	17.27	61.29		1.63	4.54	14.90	9.25	5.37
Papua New Guinea	0.13	0.08		0.17	0.11		27.27	13.19		34.77	16.82		3.26			3.97	1.63	
Philippines	1.41	0.78	2.31	3.42	1.88	5.61	20.20	9.00	22.38	49.00	21.83	54.29	5.72	3.61	7.76	6.90	2.66	4.30
Singapore	0.04	0.06	0.17	0.06	0.08	0.24	11.76	13.26	30.34	16.59	18.70	42.81	25.51	54.79	173.10	0.24	0.21	0.29
Thailand	4.01	3.00	5.98	9.88	7.40	14.74	67.38	45.87	86.83	166.08	113.07	214.03	19.50	10.67	15.97	11.30	5.42	6.46
Tonga[b]	0.00	0.01	0.01	0.00	0.01	0.01	8.63	66.41	52.55	10.27	79.05	62.54	1.12	8.97	7.29	0.80	3.42	3.13
Vanuatu	0.00	0.00		0.00	0.00		26.27	22.15		23.38	19.72		5.45	3.39		3.13	4.51	
Viet Nam[d]	0.80	1.56	1.50	2.46	4.77	4.57	10.69	18.52	16.54	32.67	56.62	50.56	6.15	8.63	5.45	8.20	5.78	8.37
SOUTH ASIA	**6.11**	**10.55**	**23.97**	**18.78**	**32.51**	**73.87**	**4.92**	**6.97**	**15.58**	**15.12**	**21.49**	**48.02**	**2.86**	**3.76**	**5.18**	**4.42**	**4.40**	**5.37**
Afghanistan[a]		0.08	0.24		0.22	0.64		3.30	7.09		8.89	19.11		2.54	5.15		4.68	1.91
Bangladesh	0.37	0.44	1.14	1.18	1.42	3.67	3.11	3.09	7.02	9.97	9.89	22.50	2.54	2.76	4.83	4.93	4.96	5.74
Bhutan	0.04	0.04	0.10	0.11	0.10	0.27	76.64	53.58	121.05	212.20	148.35	335.17	23.75	14.69	24.58	19.69	10.97	11.80
India	4.94	9.30	21.22	15.25	28.72	65.53	5.14	8.13	16.02	15.88	25.10	49.49	3.15	4.43	4.99	5.26	4.98	5.26
Maldives	0.05	0.01		0.09	0.01		200.52	22.44		343.38	38.43		38.51	5.36		12.07	1.07	
Nepal	0.15	0.18	0.56	0.46	0.54	1.70	7.17	6.94	19.47	21.54	20.87	58.53	4.08	3.67	8.73	9.64	8.23	9.53
Pakistan[b]	0.12	0.08	0.95	0.43	0.29	3.38	0.99	0.53	5.14	3.51	1.88	18.24	0.45	0.23	1.68	0.46	0.28	1.90
Sri Lanka	0.44	0.42	0.95	1.26	1.21	2.71	24.18	21.73	44.62	69.17	62.16	127.63	6.68	8.12	14.90	5.28	4.04	5.65
EUROPE AND CENTRAL ASIA	**1.85**	**12.88**	**13.62**	**5.17**	**24.80**	**25.21**	**9.41**	**50.90**	**56.76**	**26.23**	**98.04**	**105.04**	**2.94**	**12.15**	**10.65**	**0.54**	**1.90**	**1.42**
Albania	0.07	0.06	0.12	0.16	0.14	0.28	22.33	20.45	41.98	51.36	47.04	96.57	2.16	3.26	4.27	3.95	2.14	2.95
Azerbaijan[a]	0.20	0.24	0.76	0.43	0.52	1.67	25.66	28.17	79.03	56.22	61.73	173.16	6.73	8.42	16.73	8.01	4.63	2.79
Belarus	0.42	1.90	1.46	1.10	5.01	3.85	41.00	196.77	154.00	107.96	518.10	405.49	12.56	54.24	35.06	5.96	9.51	5.97
Bulgaria	0.03	0.45	0.37	0.05	0.90	0.75	3.09	58.54	52.17	6.21	117.44	104.66	0.56	12.97	14.50	0.25	2.56	1.68
Georgia	0.00	0.04	0.12	0.00	0.09	0.23	0.24	10.40	31.96	0.47	20.42	62.76	0.04	2.82	8.75	0.50	1.81	2.27
Kazakhstan		0.97	1.86		1.78	3.41		64.24	104.68		117.49	191.44		11.53	17.92		3.67	4.07
Kyrgyzstan	0.03	0.04	0.04	0.08	0.10	0.10	6.41	7.76	6.33	16.65	20.17	16.44	2.41	2.99	3.87	3.54	4.50	1.35
Latvia[a]	0.21	0.83	0.20	0.21	0.84	0.20	83.08	372.06	98.86	83.54	374.10	99.40	12.76	56.07	11.85	2.83	6.25	1.17
Lithuania[a]	0.42	0.12	0.14	2.30	0.68	0.75	115.46	37.04	46.89	633.91	203.33	257.46	68.60	25.38	27.67	29.69	3.22	2.72
Moldova	0.02	0.10	0.08	0.05	0.22	0.17	6.07	28.88	23.17	12.87	61.26	49.15	1.69	11.29	8.18	1.38	4.98	2.77
Russian Federation	0.46	6.85	9.22	0.78	11.60	15.62	3.12	47.73	63.90	5.28	80.85	108.24	0.62	9.83	10.44	0.15	1.32	1.16
Serbia[b]			0.50			1.00			68.97			139.24			14.33			2.30
Ukraine		1.26	0.34		2.93	0.79		26.83	7.60		62.24	17.64		9.38	2.09		1.97	0.60
MIDDLE EAST AND NORTH AFRICA	**8.28**	**16.48**	**11.60**	**20.43**	**36.97**	**20.17**	**29.82**	**49.88**	**122.18**	**73.58**	**111.89**	**212.35**	**5.91**	**10.77**	**16.48**	**2.44**	**3.27**	**2.68**
Algeria	0.85	2.01		2.04	4.82		29.55	60.52		70.67	144.71		7.97	15.26		2.42	4.23	

Table 2 continued

Region/country	Agricultural expenditure (billions 2011 constant US dollars)			Agricultural expenditure (billions 2011 PPP dollars)			Per capita agricultural expenditure (2011 constant US dollars)			Per capita agricultural expenditure (2011 PPP dollars)			Ratio of agricultural expenditure to agricultural GDP (%)			Share of agriculture in total expenditure (%)		
	1995	2005	2016	1995	2005	2016	1995	2005	2016	1995	2005	2016	1995	2005	2016	1995	2005	2016
Bahrain	0.02	0.02		0.04	0.03		34.98	21.37		62.41	38.13		16.86			0.54	0.41	
Egypt	1.62	1.74		5.93	6.37		25.50	22.73		93.09	82.99		9.42	7.42		4.39	3.33	
Iran	2.78	5.40		6.34	12.31		45.90	76.72		104.62	174.87		6.67	17.22		4.22	5.08	
Jordan	0.18	0.16	0.07	0.44	0.39	0.17	39.67	28.03	7.21	96.01	67.83	17.45	37.78	27.94	5.49	4.46	1.95	0.72
Kuwait[a]	0.24	0.27	0.51	0.39	0.43	0.81	149.34	118.51	128.74	239.64	190.16	206.58			56.09	0.59	0.66	0.51
Lebanon	0.03	0.02	0.05	0.05	0.04	0.10	9.24	6.15	8.91	16.61	11.05	16.01	1.97	2.53	3.50	0.40	0.31	0.42
Morocco	0.57	0.51		1.25	1.13		21.01	16.82		46.22	37.01		8.67	5.76		4.21	2.09	
Oman[c]	0.24	0.13	0.20	0.49	0.26	0.40	110.43	51.85	53.20	221.65	104.07	106.77	20.84	15.53	19.57	1.56	0.72	0.57
Qatar		0.37			0.56			432.73			651.25			445.08			1.89	
Syrian Arab Republic													9.03	8.37		10.24	6.06	
Tunisia[d]	0.63	0.57	0.74	1.49	1.36	1.76	68.97	56.42	67.97	163.96	134.13	161.59	23.53	16.62	17.09	8.17	5.22	4.16
Turkey	0.95	5.04	11.48	1.65	8.74	19.91	16.31	74.21	144.41	28.28	128.65	250.36	1.46	8.52	17.07	1.03	3.54	2.79
United Arab Emirates[c]	0.09	0.07	0.09	0.12	0.10	0.13	34.78	15.47	9.85	50.21	22.32	14.21	2.78	1.74	3.58	0.67	0.69	0.09
Yemen	0.07	0.15		0.20	0.42		4.52	7.25		12.76	20.44		1.77	4.67		1.68	1.43	
LATIN AMERICA AND THE CARIBBEAN	**21.61**	**18.53**	**14.54**	**27.04**	**24.81**	**17.82**	**55.58**	**37.93**	**42.80**	**69.55**	**50.80**	**52.47**	**10.23**	**7.98**	**6.72**	**4.14**	**2.05**	**1.59**
Argentina	0.31	0.33	0.60	0.48	0.51	0.93	8.87	8.51	13.68	13.68	13.12	21.10	1.84	1.06	1.79	0.58	0.58	0.41
Bahamas[a]	0.02	0.02	0.01	0.03	0.02	0.01	85.94	47.75	28.33	90.53	50.30	29.84	11.32	14.58	12.84	1.67	1.26	0.62
Barbados	0.03	0.03		0.03	0.03		121.67	105.27		120.65	104.39		28.30	41.57		2.80	2.17	
Belize	0.01			0.01			40.57			70.57			7.13			4.61		
Bolivia	0.01	0.07		0.01	0.15		0.83	7.16		1.95	16.87		0.32	3.05		0.35	1.39	
Brazil	16.40	10.93	11.72	18.65	12.43	13.33	101.03	58.48	56.45	114.88	66.50	64.20	20.70	11.62	9.33	5.70	2.55	1.86
Chile	0.26	0.56	1.08	0.36	0.78	1.50	18.20	34.54	60.25	25.30	48.00	83.74	3.35	6.82	9.20	1.18	1.52	1.57
Colombia[a]	0.47	0.35	0.79	0.75	0.56	1.25	12.57	8.15	16.31	20.00	12.97	25.94	1.69	1.81	3.31	1.77	0.32	0.90
Costa Rica	0.16	0.10	0.20	0.23	0.15	0.29	44.24	24.52	41.52	64.51	35.76	60.55	5.62	3.74	7.68	3.15	1.33	1.28
Dominican Republic	0.24	0.17	0.24	0.47	0.33	0.48	30.28	18.32	22.88	59.53	36.01	44.97	9.31	5.54	5.50	7.83	2.53	1.93
Ecuador[c]		0.25	0.32		0.47	0.61		17.97	20.44		34.16	38.85		4.19	4.15		1.95	1.40
El Salvador	0.04	0.05	0.05	0.07	0.10	0.09	6.64	8.47	7.29	13.20	16.84	14.50	2.13	4.60	3.40	1.69	1.15	0.80
Grenada	0.01			0.02			102.99			155.95			30.53			9.65		
Guatemala	0.07	0.17	0.11	0.14	0.35	0.23	6.31	12.62	6.38	13.54	27.10	13.69	1.00	3.49	1.86	2.72	3.15	1.54
Jamaica[a]	0.09	0.07	0.09	0.14	0.11	0.15	34.76	25.43	32.43	55.16	40.35	51.46	7.60	9.25	10.04	2.13	1.49	2.19
Mexico	3.24	4.77	5.70	5.25	7.73	9.22	34.45	44.00	48.56	55.78	71.24	78.62	9.64	14.67	15.55	3.36	2.58	2.32
Panama[c]	0.05	0.08	0.12	0.08	0.15	0.21	16.93	23.91	30.23	30.94	43.68	55.24	5.15	6.08	9.68	1.64	1.76	1.35
Paraguay																		
Peru		0.24	0.54		0.44	0.97		8.75	16.90		15.85	30.59		3.04	3.61		1.39	1.91
St. Vincent and the Grenadines	0.00	0.00		0.01	0.01		34.98	36.64		55.85	58.49		9.17	11.47		3.46	2.33	
Trinidad and Tobago[b]	0.13	0.18	0.08	0.20	0.29	0.14	99.80	136.78	61.87	162.43	222.63	100.70	54.56	174.04	92.55	4.49	3.38	0.86
Uruguay	0.09	0.16		0.11	0.21		27.39	49.11		34.61	62.07		3.49	5.47		1.04	1.72	
AFRICA SOUTH OF THE SAHARA	**3.59**	**6.66**	**9.50**	**7.84**	**14.44**	**19.48**	**7.44**	**10.14**	**12.62**	**16.27**	**21.97**	**25.86**	**3.15**	**3.75**	**3.85**	**2.74**	**3.16**	**2.19**
Angola	0.12	0.35	0.28	0.17	0.49	0.38	8.48	18.07	9.66	11.66	24.85	13.29	7.11			1.74	2.00	0.98

Table 2 continued

Region/country	Agricultural expenditure (billions 2011 constant US dollars)			Agricultural expenditure (billions 2011 PPP dollars)			Per capita agricultural expenditure (2011 constant US dollars)			Per capita agricultural expenditure (2011 PPP dollars)			Ratio of agricultural expenditure to agricultural GDP (%)			Share of agriculture in total expenditure (%)		
	1995	2005	2016	1995	2005	2016	1995	2005	2016	1995	2005	2016	1995	2005	2016	1995	2005	2016
Benin	0.06	0.07	0.23	0.13	0.15	0.50	10.09	8.34	20.84	22.24	18.38	45.95	4.44	4.28	9.85	7.26	5.47	11.78
Botswana	0.17	0.18	0.31	0.30	0.34	0.57	106.06	99.48	139.91	192.68	180.72	254.18	46.66	85.33	78.86	5.96	4.49	5.07
Burkina Faso	0.39	0.15	0.29	0.86	0.34	0.64	38.67	11.46	15.50	85.41	25.30	34.24	28.75	5.64	7.19	45.68	9.01	8.18
Burundi[c]	0.03	0.00	0.02	0.08	0.01	0.05	4.58	0.28	1.92	13.57	0.84	5.67	3.71	0.28	1.98	5.10	0.33	2.79
Cabo Verde[d]	0.00	0.02	0.03	0.00	0.03	0.05	5.04	36.25	65.28	8.22	59.15	106.50	2.48	14.15	21.06		3.32	4.93
Cameroon													2.45	4.17		4.16	3.76	
Central African Republic[c]	0.02	0.01	0.02	0.04	0.01	0.04	7.05	1.57	4.76	13.00	2.89	8.77	3.47	0.68	3.38	9.94	2.04	9.91
Congo, Rep.[d]	0.01	0.05	0.12	0.01	0.07	0.19	3.15	12.15	25.68	5.13	19.82	41.88	1.08	9.19	20.25	0.34	1.90	2.20
Côte d'Ivoire	0.17	0.10	0.48	0.36	0.22	1.00	12.01	5.69	20.38	24.83	11.76	42.13	3.46	1.94	5.88	3.56	2.34	4.52
Democratic Rep. of Congo[d]	0.00	0.03	0.42	0.00	0.05	0.74	0.07	0.49	6.06	0.12	0.86	10.68	0.03	0.67	7.39	0.18	0.70	6.16
Equatorial Guinea	0.01	0.04		0.01	0.06		14.68	52.52		23.52	84.13							
eSwatini	0.04	0.06	0.09	0.07	0.11	0.16	36.60	55.13	63.90	68.13	102.63	118.97	12.06	13.87	16.29	5.68	5.56	4.07
Ethiopia	0.16	0.63	0.88	0.54	2.16	3.01	2.74	8.20	8.55	9.40	28.18	29.37	3.15	9.34	5.24	9.72	15.93	9.23
Gambia	0.01	0.01	0.01	0.02	0.02	0.03	4.85	4.48	5.59	14.36	13.29	16.57	4.68	3.20	6.27	7.88	6.91	3.20
Ghana	0.04	0.04	0.11	0.09	0.09	0.24	2.45	1.96	3.91	5.30	4.23	8.45	0.66	0.44	1.20	0.73	0.53	0.92
Guinea-Bissau[c]	0.00	0.00	0.00	0.00	0.00	0.00	0.40	1.05	0.83	0.86	2.26	1.79	0.10	0.38	0.29	1.19	7.77	0.88
Kenya	0.45	0.24	0.31	1.18	0.61	0.80	16.59	6.58	6.34	42.97	17.05	16.41	7.37	3.16	1.75	7.00	3.87	1.77
Lesotho	0.09	0.03	0.06	0.16	0.06	0.11	50.33	16.35	25.93	93.16	30.25	48.00	65.16	27.23	33.75	12.41	3.80	3.75
Liberia[d]	0.00	0.00	0.00	0.02	0.00	0.02	0.06	0.01	0.03	8.04	1.10	4.27	3.97	0.28	1.43	2.76	1.34	1.97
Madagascar	0.07	0.17	0.01	0.20	0.52	0.04	4.94	9.36	0.57	14.86	28.14	1.70	4.28	7.89	0.57	6.10	9.49	1.17
Malawi	0.10	0.08	0.49	0.21	0.17	1.01	10.32	6.20	27.12	21.18	12.73	55.67	9.61	4.62	19.84	8.85	6.86	17.43
Mali	0.22	0.38	0.40	0.50	0.86	0.89	23.02	29.92	22.11	51.69	67.16	49.65	10.47	11.74	6.57	17.28	18.27	12.31
Mauritius	0.07	0.08	0.08	0.13	0.14	0.14	63.56	62.15	60.96	114.46	111.92	109.77	14.07	17.03	17.60	5.86	3.70	1.93
Mozambique		0.15	0.39		0.27	0.72		7.04	13.68		12.76	24.81		7.33	9.78		7.66	7.26
Namibia	0.12	0.14	0.26	0.19	0.21	0.41	71.99	67.91	105.02	112.09	105.75	163.53	19.94	14.03	27.29	6.04	5.14	4.45
Niger[b]	0.07	0.19	0.36	0.16	0.42	0.77	7.76	14.29	18.75	16.56	30.50	40.03	5.26	16.17	12.09	13.17	20.02	12.30
Nigeria	0.44	1.82	0.17	0.92	3.77	0.36	4.10	13.10	0.94	8.49	27.11	1.94	1.00	2.02	0.17	3.60	4.99	0.63
Rwanda		0.05	0.27		0.11	0.61		5.08	22.29		11.71	51.32		2.92	9.76		4.85	10.55
Senegal	0.07	0.22	0.51	0.14	0.44	1.03	8.24	19.45	33.31	16.46	38.84	66.53	4.92	12.61	18.44	5.23	7.85	10.60
Seychelles	0.01	0.01	0.01	0.01	0.02	0.02	85.62	122.24	137.73	158.45	226.22	254.87	25.15	39.02	49.48	1.99	3.05	2.61
Sierra Leone[e]	0.00	0.01	0.05	0.01	0.02	0.14	0.87	1.50	7.35	2.44	4.20	20.57	0.64	0.80	2.60	1.57	0.90	5.93
South Africa	0.35	0.72	2.86	0.53	1.09	4.36	8.22	14.72	51.14	12.51	22.39	77.78	3.91	8.69	29.06	0.51	0.82	1.10
Sudan													0.05	4.74		3.59	6.77	
Tanzania, United Rep.	0.19	0.24	0.27	0.58	0.71	0.79	6.49	6.04	4.77	19.35	18.01	14.22	3.33	3.56	1.94	8.55	6.26	2.98
Togo	0.03	0.02	0.08	0.07	0.05	0.18	7.56	3.73	10.87	16.58	8.19	23.86	3.65	1.76	4.14	6.13	3.52	7.39
Uganda	0.01	0.07	0.17	0.04	0.21	0.51	0.58	2.48	4.04	1.77	7.50	12.23	0.43	2.45	3.05	1.87	3.06	3.91
Zambia	0.06	0.31	0.48	0.11	0.63	0.98	6.16	25.67	29.04	12.59	52.46	59.35	4.40	14.49	26.09	2.80	8.30	6.76
HIGH-INCOME EUROPEAN COUNTRIES	**102.33**	**73.59**	**49.67**	**95.00**	**70.95**	**45.48**	**215.08**	**148.58**	**102.85**	**199.66**	**143.26**	**94.17**	**25.85**	**24.65**	**17.84**	**1.44**	**0.89**	**0.53**
Austria	3.94	2.22	1.60	3.41	1.92	1.38	495.73	269.40	182.82	428.94	233.10	158.19	58.80	44.83	32.30	2.32	1.10	0.70

Table 2 continued

Region/country	Agricultural expenditure (billions 2011 constant US dollars) 1995	2005	2016	Agricultural expenditure (billions 2011 PPP dollars) 1995	2005	2016	Per capita agricultural expenditure (2011 constant US dollars) 1995	2005	2016	Per capita agricultural expenditure (2011 PPP dollars) 1995	2005	2016	Ratio of agricultural expenditure to agricultural GDP (%) 1995	2005	2016	Share of agriculture in total expenditure (%) 1995	2005	2016
Belgium	0.65	0.33	0.35	0.56	0.28	0.30	64.32	31.15	30.46	55.62	26.94	26.33	13.11	8.09	10.13	0.33	0.13	0.12
Croatia	0.23	0.47	0.42	0.33	0.67	0.60	49.17	106.05	100.51	70.01	151.02	143.14	9.26	18.25	19.16	2.50	2.33	1.39
Cyprus	0.24	0.30	0.20	0.25	0.31	0.20	280.89	296.20	169.06	289.23	305.00	174.07	30.07	46.03	42.04	4.61	3.27	1.98
Czech Republic[b]	4.08	1.13	0.95	5.41	1.50	1.26	395.37	110.66	90.61	524.26	146.74	120.14	68.66	25.73	16.63	5.27	1.36	0.98
Denmark	0.55	0.59	0.55	0.40	0.43	0.40	105.15	109.40	96.32	75.60	78.66	69.26	7.12	15.41	18.80	0.35	0.34	0.28
Estonia	0.07	0.12	0.12	0.10	0.17	0.17	48.84	89.29	89.26	68.67	125.56	125.51	11.95	17.76	20.25	1.47	1.62	1.10
Finland	8.33	2.94	2.13	6.67	2.35	1.70	1,630.82	560.26	386.87	1,306.29	448.77	309.89	124.67	50.27	33.78	7.75	2.33	1.40
France	10.40	11.26	5.79	8.89	9.63	4.95	174.71	178.27	86.54	149.38	152.42	73.99	19.76	24.85	13.40	0.89	0.79	0.34
Germany	17.93	8.76	7.25	16.36	7.99	6.61	219.56	106.21	88.01	200.25	96.86	80.27	62.86	37.42	32.66	1.09	0.56	0.41
Greece	3.05	0.08	0.65	3.07	0.08	0.66	288.42	6.99	60.77	290.93	7.05	61.29	18.56	0.56	7.32	3.11	0.06	0.51
Hungary	1.89	1.76	0.74	3.06	2.86	1.20	183.35	174.97	75.30	296.64	283.07	121.82	27.29	34.31	12.74	3.51	2.56	1.02
Iceland		0.17	0.14		0.14	0.12		562.80	414.27		482.86	355.43		25.36	15.29	7.22	2.95	1.73
Ireland	1.21	1.28	0.98	1.05	1.11	0.85	335.75	308.30	206.68	290.46	266.72	178.81	19.22	55.45	30.58	2.71	1.72	1.05
Italy	9.17	8.77	5.75	8.69	8.32	5.46	161.23	151.35	94.91	152.87	143.50	89.99	15.57	18.81	13.72	0.89	0.81	0.53
Luxembourg	0.19	0.22	0.20	0.15	0.18	0.16	466.51	480.40	341.59	370.77	381.81	271.49	62.15	121.25	117.99	1.44	1.01	0.68
Malta	0.01	0.09	0.05	0.02	0.11	0.07	35.32	222.44	115.53	44.25	278.70	144.75	8.97	54.62	35.54	0.53	2.50	1.11
Netherlands	2.02	1.67	1.01	1.73	1.43	0.87	130.40	102.13	59.11	112.19	87.87	50.86	10.52	11.28	6.55	0.60	0.48	0.25
Norway	5.28	3.20	3.27	3.26	1.97	2.01	1211.33	692.21	623.78	747.46	427.14	384.91	57.67	48.39	28.12	2.99	1.62	1.18
Poland	2.69	2.91	2.48	4.42	4.79	4.07	69.66	76.30	65.23	114.57	125.49	107.28	18.32	25.19	17.31	2.68	1.65	1.00
Portugal	1.73	1.21	0.59	2.00	1.40	0.68	172.54	115.31	56.83	199.20	133.12	65.61	19.03	21.81	12.75	2.14	1.07	0.54
Romania[b]	2.79	2.01	1.61	5.49	3.96	3.16	123.11	94.50	80.73	242.14	185.86	158.77	12.78	15.21	17.08	6.77	3.81	2.32
Slovakia		0.26	0.34		0.37	0.48		49.06	62.49		69.69	88.77		10.81	9.14		0.88	0.73
Slovenia	0.50	0.32	0.24	0.58	0.37	0.28	251.21	158.64	116.97	289.61	182.89	134.85	42.69	29.69	23.95	3.61	1.51	1.00
Spain	6.28	9.06	5.13	6.32	9.13	5.16	158.04	207.62	110.28	159.20	209.14	111.09	16.34	23.44	13.28	1.43	1.66	0.79
Sweden	2.51	1.14	0.83	1.84	0.84	0.61	284.07	126.56	84.00	208.57	92.92	61.68	27.39	22.45	11.53	1.06	0.43	0.27
Switzerland	15.03	5.42	5.45	9.55	3.45	3.47	2,134.22	729.33	651.39	1,356.19	463.46	413.93	198.78	101.97	110.46	9.24	2.59	2.10
United Kingdom	1.56	5.87	3.43	1.38	5.19	3.03	26.94	97.18	52.32	23.82	85.90	46.25	6.40	39.94	21.91	0.22	0.56	0.28
OTHER HIGH-INCOME COUNTRIES	**44.44**	**62.04**	**26.51**	**41.20**	**58.71**	**24.74**	**90.05**	**115.92**	**74.43**	**83.50**	**109.69**	**69.47**	**17.95**	**19.05**	**12.09**	**1.25**	**1.32**	**0.58**
Australia	3.06	3.71	4.78	1.96	2.38	3.06	169.45	181.89	197.23	108.72	116.70	126.54	11.83	10.39	11.70	1.24	1.14	0.76
Canada[b]	4.78	5.13	6.44	3.81	4.10	5.14	162.70	158.86	181.13	129.84	126.78	144.55			23.55	1.93	2.14	0.96
Israel	0.90	0.35	0.62	0.82	0.32	0.56	162.51	50.46	72.29	147.41	45.77	65.57	35.87	10.95	17.17	1.49	0.38	0.52
Japan[b]	13.69	13.31	7.13	10.17	9.88	5.30	109.12	104.15	56.06	81.04	77.35	41.63	14.79	19.27	10.55	1.69	1.33	0.60
Rep. of Korea[b]	8.39	8.82	10.12	10.88	11.44	13.12	185.99	183.15	199.40	241.20	237.52	258.60	27.27	32.80	36.38	10.02	4.74	2.37
New Zealand[d]	0.36	0.88	1.03	0.31	0.75	0.88	97.48	213.98	234.67	83.04	182.29	199.91	4.38	10.34	8.28	1.06	1.38	1.32
United States	13.26	29.84	21.12	13.26	29.84	21.12	49.81	100.96	65.30	49.81	100.96	65.30		17.83	12.05	0.64	1.07	0.56

Note: PPP (purchasing power parity) dollars measure the relative purchasing power of currencies across countries by eliminating national differences in pricing levels for a wide range of goods and services.

a = last year of data available is 2015; b = last year of data available is 2014; c = last year of data available is 2013; d = last year of data available is 2012; e = last year of data available is 2013, with the exception of the estimated share of agriculture in total expenditure, which is for 2012.

Regional totals and averages are calculated for the countries for which observations are available in the years 1995, 2005, and 2016. Averages are weighted by the denominator of the indicator being used (population, agricultural GDP, or total spending).

Food Policy Research Capacity Indicators (FPRCI)

Food policy research plays a crucial role in guiding the agricultural transformation of developing countries. To achieve food security goals, countries need to strengthen their capacity to conduct food policy research. Strong local policy research institutions help in shaping an evidence-based policymaking process. Measuring national capacity for food policy research is important for identifying capacity gaps in food policy research and guiding the allocation of resources to fill those gaps.

"Food policy research capacity" is defined as any socioeconomic or policy-related research capacity in the area of food, agriculture, nutrition, or natural resources. To measure this capacity, the International Food Policy Research Institute (IFPRI) developed a set of indicators of the quantity and quality of policy research at the country level.

INDICATORS

IFPRI created the FPRCI database in 2010 and has continued to expand and refine it. The data presented in Table 3 are currently collected for 33 countries; data for Myanmar was added in 2017. A consistent methodology is followed to enable comparison of values across time and countries. The database was most recently updated with numbers for 2018.

Analysts/researchers is a head count of professionals employed at local organizations whose work involves food policy research or analysis. To introduce some uniformity, IFPRI also presents a modified quantification of the headcount: full-time equivalent analysts/researchers with PhD. To obtain an indicator of per capita food policy research capacity, this research capacity is then divided by the country's rural population (full-time equivalent researchers per million rural residents). This helps to illustrate the level of effort invested in local food policy research in a country. This indicator was last updated in 2015.

The quality of a country's food policy research capacity is estimated by tallying the number of relevant international publications in peer-reviewed journals. IFPRI views this as a reflection of the local enabling environment for food policy research. This indicator allows for comparison across countries, as it ensures an internationally accepted standard of quality for publications. The final indicator is derived by dividing the number of international publications by the number of full-time equivalent researchers with a PhD, providing a measure of productivity.

TRENDS IN FOOD POLICY RESEARCH

Overall food policy research capacity across all countries did not change from last year's level, but there was substantial variation across countries and regions. For instance, the number of food policy research publications in Colombia had been steadily declining until 2017 but increased in 2018. Other countries, including Nepal, Bangladesh, and Ghana have seen a steady increase since 2014.

Most Asian countries showed little change in number of publications, with the exception of Bangladesh, Lao PDR, and Nepal, which saw an increase in publications in 2018. In Africa, Ethiopia, Ghana, Malawi, Mali, and Tanzania saw an increase in the number of publications. In Latin America, apart from Guatemala, all countries experienced either a slight increase or no change in the number of publications. A few countries, notably South Africa, Indonesia, and Kenya, saw a decrease in the number of publications produced.

To better understand long-term trends in food policy research, we calculated five-year moving averages of the number of publications for the period 2009–2018. For the Asia region, except China, the moving average for most countries remained unchanged. In Africa, the moving average for all countries increased or remained the same, with the exception of Liberia, Madagascar, South Africa, and Swaziland (eSwatini).

IFPRI will continue to update and expand this database to include additional countries to better facilitate cross-country comparisons. This will also facilitate identification of the minimum food policy research capacity threshold for a country. It is hoped that such data will aid in informing national policymakers of the importance of investing in local food policy research capacity. This data will also provide donors with a framework for prioritizing investments to strengthen food policy research capacity across as well as within countries.

DOWNLOAD DATA
https://doi.org/10.7910/DVN/QU15AW

CONTACT
Suresh Babu (s.babu@cgiar.org)
and Paul Dorosh (p.dorosh@cgiar.org)

TABLE 3 Food policy research capacity indicators, 2011-2018

Country	Analysts/researchers (head count) in 2015	FTE analysts/ researchers with PhD in 2015	International publications produced from 2011-2018	FTE analysts/ researchers with PhD per million rural population in 2017	Publications per FTE researcher with PhD
Afghanistan	43	3.0	6	0.11	2.02
Bangladesh	66	22.9	88	0.22	3.84
Benin	38	4.3	31	0.72	7.21
Burundi	39	5.1	5	0.54	0.98
China	2,000	1,332.5	1,632	2.29	1.22
Colombia	85	6.5	33	0.67	5.12
Ethiopia	141	30.4	37	0.36	1.22
Ghana	153	23.3	93	1.81	3.99
Guatemala	45	11.9	4	1.43	0.34
Honduras	33	6.1	6	1.52	0.98
Indonesia	146	42.4	25	0.35	0.59
Kenya	155	31.6	88	0.87	2.78
Lao PDR	9	1.8	13	0.39	7.43
Liberia	34	3.1	2	1.32	0.65
Madagascar	187	11.5	11	0.71	0.95
Malawi	68	18.2	38	1.17	2.09
Mali	60	10.1	13	0.93	1.29
Mozambique	37	3.3	15	0.17	4.51
Myanmar	97	46.5	10	1.25	0.22
Nepal	27	3.7	26	0.15	7.12
Niger	29	8.8	7	0.49	0.79
Nigeria	349	77.4	53	0.80	0.68
Peru	54	7.2	24	1.00	3.36
Rwanda	64	5.5	12	0.54	2.18
Senegal	71	9.3	19	1.10	2.04
South Africa	198	50.3	268	2.60	5.33
Swaziland* (eSwatini)	32	2.9	1	0.27	0.35
Tanzania, United Rep.	91	20.8	51	0.54	2.46
Togo	81	6.8	5	1.49	0.73
Uganda	34	10.9	33	0.33	3.02
Viet Nam	175	32.5	17	0.53	0.52
Zambia	29	5.3	22	0.54	4.15
Zimbabwe	42	8.9	18	0.79	2.03

Note: * = rural population from 2016 is used. FTE = full-time equivalent.

Agricultural Total Factor Productivity (TFP)

Increasing the efficiency of agricultural production—getting more output from the same amount of resources—is critical for improving food security. To measure the efficiency of agricultural systems, we use total factor productivity (TFP). TFP is an indicator of how efficiently agricultural land, labor, capital, and materials (agricultural inputs) are used to produce a country's crops and livestock (agricultural output). It is calculated as the ratio of total agricultural output to total production inputs. When more output is produced from a constant amount of resources, meaning that resources are being used more efficiently, TFP increases. Measures of land and labor productivity—partial factor productivity (PFP) measures—are calculated as the ratio of total output to total agricultural area (land productivity) and to the number of economically active persons in agriculture (labor productivity). Because PFP measures are easy to estimate, they are often used to measure agricultural production performance. These measures normally show higher rates of growth than TFP, because growth in land and labor productivity can result not only from increases in TFP but also from a more intensive use of other inputs (such as fertilizer or machinery). Indicators of both TFP and PFP contribute to the understanding of agricultural systems needed for policy and investment decisions by enabling comparisons across time and across countries and regions.

TRENDS IN PRODUCTIVITY

Table 4 presents estimates of TFP and land and labor productivity measures for developing countries and regions for three periods between 1991 and 2015. It uses the most recent data on outputs and inputs from the Economic Research Service of the United States Department of Agriculture (ERS-USDA), an internationally consistent and comparable dataset on production and input quantities built using data from the FAOSTAT database of the Food and Agriculture Organization of the United Nations (FAO), supplemented with data from national statistical sources.[1]

Average TFP growth for all developing regions in 2011–2015 shows lower values than average growth rates of the previous decade (2001–2010). Sluggish TFP growth in Africa south of the Sahara in 2001–2010 decreased to 0.4 percent in 2011–2015. TFP growth in Latin America and the Caribbean is the highest among developing regions in recent years at 1.9 percent in 2011–2015, but below the 2.3 percent growth in the region in 2001–2010, largely because of slower growth in Brazil (1.6 percent in 2011–2015 compared with 3.3 percent in 2001–2010). Growth in Asia and the Pacific and the Middle East and North Africa for 2011–2015 is similar to growth of these regions in 2001–2010, with growth in China slowing to less than 2 percent, and India's growth accelerating from 1.1 percent in 2001–2010 to 1.7 for 2011–2015.

DATA AND METHODOLOGY

The output values from the ERS-USDA dataset used to estimate TFP are gross agricultural outputs. Inputs are agricultural land, in hectares of "rainfed cropland equivalents," measured as the sum of rainfed cropland (weight equals 1.00), irrigated cropland (weight varies from 1.00 to 3.00 depending on region), and permanent pasture (weight varies from 0.02 to 0.09 depending on region); a measure of labor is approximated using the number of economically active persons in agriculture; fertilizer is expressed in tons of fertilizer nutrients used; total stock of farm machinery is measured as the number of "40-CV tractor equivalents"; total livestock capital on farms is in "cattle equivalents," calculated using FAO estimates of animal stocks on farms, with species weighted by their respective size (nondairy cattle taking a weight of 1.00). Finally, total animal feed from crops and crop processing residues is measured in "dry-matter equivalents."

Land and labor productivity measures for the regions (such as Africa south of the Sahara) reflect a

DOWNLOAD DATA
https://doi.org/10.7910/DVN/9IOAKR

CONTACT
Alejandro Nin-Pratt (a.ninpratt@cgiar.org)

weighted average of individual country productivity measures using average outputs (1991–2014) of each country as weights. TFP is calculated using a growth accounting approach. This approach defines TFP as the ratio of an output index and an input index. As input prices are not available, input shadow prices were obtained for each country and year by estimating the parameters of distance functions through linear programming.

TABLE 4 Average annual growth of agricultural output and total factor productivity (TFP), and levels of land and labor productivity, various years

Region/country	Land productivity (in constant 2004-2006 US dollars)				Labor productivity (in constant 2004-2006 US dollars)				Output growth (%)			TFP growth (%)		
	1990	2000	2010	2015	1990	2000	2010	2015	1991-2000	2001-2010	2011-2015	1991-2000	2001-2010	2011-2015
AFRICA SOUTH OF THE SAHARA	**190**	**239**	**319**	**335**	**1,084**	**1,438**	**1,887**	**2,078**	**4.1**	**3.7**	**2.1**	**1.5**	**1.0**	**0.4**
Angola	15	24	62	67	252	315	605	567	4.9	9.8	1.6	0.5	4.6	0.0
Benin	471	582	656	669	780	1,109	1,243	1,485	6.0	2.7	4.4	1.8	0.9	3.3
Botswana	9	12	15	14	1,071	1,090	1,264	1,114	3.2	2.5	-1.5	3.1	1.8	5.8
Burkina Faso	111	145	196	207	297	294	387	335	2.8	6.0	0.4	-0.5	1.4	-1.8
Burundi	526	607	693	642	406	344	301	250	-0.4	2.3	-1.4	-0.5	0.2	-1.9
Cameroon	444	534	661	703	713	842	1,398	1,612	3.1	5.7	3.2	0.9	3.9	1.4
Cabo Verde	256	448	543	440	816	1,263	1,468	1,467	4.4	0.6	-2.0	3.7	1.8	-0.4
Central African Republic	148	197	228	235	526	667	754	758	3.6	1.8	0.9	1.5	0.8	0.3
Chad	18	28	46	45	446	553	759	677	4.8	5.5	-0.8	1.3	2.0	0.0
Comoros	596	606	581	547	409	401	346	311	2.2	1.0	0.4	-2.6	-0.1	0.9
Congo, Dem. Rep.	179	159	160	167	493	347	293	281	-1.5	0.2	1.0	-0.4	-1.4	-1.4
Congo, Republic	20	27	39	43	466	550	761	819	2.8	3.9	2.1	0.8	0.8	1.9
Côte d'Ivoire	208	290	295	315	1,520	1,943	2,259	2,717	3.4	0.9	3.8	1.1	-0.3	0.8
Djibouti	39	29	37	42	267	201	236	250	-1.0	3.1	2.8	-2.2	1.6	3.7
Equatorial Guinea	168	181	231	249	363	277	258	244	0.0	1.4	1.1	-1.8	0.8	0.9
Ethiopia	133	146	253	303	255	218	302	324	8.8	6.5	4.1	-1.5	2.0	1.5
Gabon	41	52	60	65	949	1,245	1,550	1,595	2.4	1.4	1.7	1.1	1.6	1.5
Gambia	129	256	296	268	233	323	328	210	6.0	3.1	-5.6	2.5	0.0	-4.3
Ghana	178	312	458	536	567	904	1,134	1,191	7.5	4.8	3.4	4.4	0.5	1.0
Guinea	82	114	145	154	415	433	507	498	3.8	3.4	1.9	-1.0	0.1	1.0
Guinea-Bissau	114	160	224	236	450	609	753	742	4.7	3.7	1.8	1.6	2.5	1.7
Kenya	158	177	285	287	513	418	561	523	1.1	5.1	0.7	0.0	2.4	-1.8
Lesotho	56	77	79	75	436	528	529	464	3.1	0.2	-1.8	2.5	0.6	2.9
Liberia	108	153	156	167	457	525	443	437	3.9	0.5	1.4	0.5	-0.9	1.7
Madagascar	70	66	91	88	610	491	497	408	0.5	3.3	-0.9	1.0	1.1	-0.8
Malawi	245	408	540	513	302	495	625	530	6.4	4.6	-0.5	3.5	0.7	-3.5
Mali	44	61	95	114	822	1,052	1,359	1,501	4.3	5.1	4.5	2.8	1.7	1.7
Mauritania	9	12	15	17	767	830	747	756	3.6	1.8	2.9	0.8	-0.5	2.6
Mauritius	2,528	2,749	3,372	3,506	3,174	3,951	5,556	6,437	0.3	0.4	-0.9	-1.2	1.0	0.5
Mozambique	24	37	63	62	221	249	356	303	4.3	5.8	-0.8	0.8	1.7	0.1
Namibia	10	10	11	11	1,689	1,529	1,611	1,596	0.4	0.6	0.6	-1.9	2.5	1.6
Niger	35	49	77	87	500	545	816	747	4.2	7.4	1.9	1.2	3.5	-0.1
Nigeria	291	387	505	500	1,305	2,137	2,480	2,495	5.0	2.6	2.6	3.2	0.3	0.7

Table 4 continued

Region/country	Land productivity (in constant 2004-2006 US dollars)				Labor productivity (in constant 2004-2006 US dollars)				Output growth (%)			TFP growth (%)		
	1990	2000	2010	2015	1990	2000	2010	2015	1991-2000	2001-2010	2011-2015	1991-2000	2001-2010	2011-2015
Réunion (France)	2,061	3,206	3,203	3,023	8,487	16,648	30,201	43,703	2.0	0.8	-0.7	2.6	0.9	0.6
Rwanda	599	675	1,040	1,010	459	424	536	503	0.8	5.1	1.4	0.7	2.3	0.2
Sao Tome and Principe	289	517	572	607	512	926	752	705	7.1	0.5	1.4	4.9	-1.0	-0.6
Senegal	111	146	197	205	374	395	427	384	3.0	3.5	1.0	0.1	0.5	-0.3
Seychelles	1,773	1,648	2,701	3,375	236	264	143	155	2.2	-4.8	1.6	-0.3	1.5	6.6
Sierra Leone	147	127	297	351	400	330	858	935	-2.5	11.8	2.7	0.2	5.1	3.8
Somalia	33	33	38	39	816	707	654	600	-0.1	1.3	0.9	1.4	0.2	4.1
South Africa	96	110	135	141	5,713	7,280	10,846	12,926	1.6	1.9	0.8	3.4	1.5	0.4
Sudan, former	32	55	95	89	732	1,106	1,168	976	6.4	2.3	-2.1	3.1	-0.1	-5.5
Swaziland (eSwatini)	220	205	249	264	1,958	1,672	2,151	2,342	-1.0	1.9	1.4	-1.6	3.3	2.7
Tanzania, United Rep.	139	144	216	278	374	332	473	606	1.3	5.8	7.7	-0.4	-1.0	3.9
Togo	223	272	329	329	512	579	667	698	3.0	3.3	2.9	1.5	1.4	1.2
Uganda	384	435	447	441	584	591	455	416	2.6	0.2	1.0	-0.2	-0.3	0.5
Zambia	39	46	89	100	339	362	610	587	2.5	7.2	2.1	2.5	3.3	-2.3
Zimbabwe	127	135	108	111	551	574	497	431	1.7	-1.5	-0.4	0.3	-0.3	0.2
LATIN AMERICA AND THE CARIBBEAN	**255**	**336**	**468**	**526**	**5,710**	**7,903**	**12,404**	**15,447**	**3.2**	**3.4**	**2.5**	**1.3**	**2.3**	**1.9**
Argentina	190	251	285	329	16,647	22,096	29,888	35,954	2.8	2.7	3.1	0.2	0.9	3.4
Bahamas	1,656	1,777	2,234	2,522	3,312	4,621	6,702	8,828	1.5	3.7	1.0	0.1	3.7	3.6
Belize	725	1,054	1,062	1,088	5,076	6,544	5,377	4,974	5.4	0.6	0.9	2.6	-0.8	-1.2
Bolivia	48	65	92	113	1,388	1,509	1,719	1,909	3.6	3.7	4.2	0.6	-0.3	2.5
Brazil	253	342	514	568	4,341	6,699	12,710	16,631	3.8	4.5	2.6	1.6	3.3	1.6
Chile	279	414	519	535	4,736	6,487	8,457	8,914	3.4	2.7	0.7	0.9	2.7	-1.7
Colombia	216	255	310	351	2,907	3,184	3,720	4,609	1.6	1.4	3.5	1.6	1.2	2.7
Costa Rica	707	1,234	1,563	1,740	5,292	6,942	8,831	10,197	3.3	2.3	2.1	2.4	1.9	1.3
Cuba	694	567	442	523	5,594	5,060	4,865	6,190	-2.3	-2.6	2.7	0.3	-1.8	2.2
Dominican Republic	630	750	1,149	1,350	2,563	3,423	5,972	7,670	1.6	3.8	2.9	1.7	2.8	4.4
Ecuador	410	556	869	1,119	2,922	3,652	5,117	5,141	3.3	3.7	-0.1	0.3	2.7	-0.4
El Salvador	599	676	736	681	1,287	1,529	1,905	1,938	1.8	1.1	-0.7	0.5	1.2	-0.6
French Guiana (France)	875	960	897	900	1,671	2,008	1,817	2,347	1.8	-0.1	5.1	1.2	1.2	4.5
Guatemala	469	634	1,168	1,538	1,355	1,901	2,251	2,547	3.4	4.9	4.6	1.7	2.2	4.4
Guyana	105	185	210	284	3,138	5,650	6,656	9,346	5.5	1.1	6.0	3.1	1.1	5.7
Haiti	582	561	703	863	521	479	584	670	0.2	3.3	3.8	-1.5	2.9	4.2
Honduras	355	442	616	674	1,747	1,762	2,982	3,302	1.0	4.3	1.8	0.5	3.5	1.9
Jamaica	1,031	1,117	1,226	1,265	1,785	2,158	2,531	2,754	0.9	0.2	0.6	0.2	2.8	1.8
Lesser Antilles	1,904	2,015	1,567	1,674	3,790	4,169	3,586	3,670	-0.6	-2.5	0.0	-0.2	-0.6	0.6
Mexico	216	278	341	378	2,640	3,348	4,497	5,289	2.6	2.1	2.0	2.2	2.0	2.2
Nicaragua	162	208	299	319	1,667	2,729	4,263	4,882	4.9	3.4	1.4	3.3	3.1	0.3
Panama	383	367	403	434	3,202	3,058	3,515	3,922	0.1	1.1	1.4	-2.0	1.0	2.3
Paraguay	156	143	245	279	4,638	4,070	6,253	6,943	0.9	5.8	3.2	-2.5	1.6	0.6
Peru	156	255	366	415	1,223	1,754	2,364	2,648	5.5	4.0	2.8	2.2	2.1	1.7

Table 4 continued

Region/country	Land productivity (in constant 2004-2006 US dollars)				Labor productivity (in constant 2004-2006 US dollars)				Output growth (%)			TFP growth (%)		
	1990	2000	2010	2015	1990	2000	2010	2015	1991-2000	2001-2010	2011-2015	1991-2000	2001-2010	2011-2015
Puerto Rico (USA)	990	1,433	1,734	1,664	8,790	11,811	20,998	27,417	-2.0	-0.5	-0.4	0.4	3.2	3.6
Suriname	1,343	1,061	1,741	1,623	4,076	3,112	4,132	4,337	-2.4	3.8	1.0	-1.3	4.7	3.2
Trinidad and Tobago	1,743	2,188	2,664	2,674	2,632	2,992	3,061	3,358	0.9	-0.2	0.1	0.9	-0.2	-1.0
Uruguay	146	191	280	302	11,776	14,487	21,719	23,953	2.7	3.5	1.5	0.9	1.3	-0.1
Venezuela	196	261	299	322	4,914	6,945	8,931	10,510	2.8	1.4	1.4	2.7	0.5	1.5
ASIA AND THE PACIFIC	**646**	**909**	**1,219**	**1,355**	**803**	**1,104**	**1,738**	**2,298**	**3.8**	**3.4**	**2.5**	**1.7**	**1.7**	**1.5**
Afghanistan	54	68	93	96	823	643	648	599	2.2	3.1	0.6	2.5	-1.6	-0.6
Armenia	560	509	566	854	2,523	1,188	2,146	3,779	-0.6	3.8	7.6	4.6	2.7	5.6
Azerbaijan	558	335	528	627	2,058	1,051	1,522	1,760	-4.0	4.6	3.4	2.6	0.2	0.7
Bangladesh	1,073	1,610	2,332	2,583	355	466	661	747	3.1	3.5	1.9	-0.5	0.7	0.7
Bhutan	229	306	377	387	650	959	636	582	4.4	1.9	0.8	4.3	0.4	-0.9
Cambodia	275	406	698	836	411	489	755	834	4.6	6.8	3.6	2.4	1.7	-0.2
China	433	691	970	1,093	582	1,053	1,921	2,788	5.2	3.5	2.6	2.8	2.2	1.9
Fiji	608	542	443	464	2,150	1,855	1,493	1,529	-0.7	-2.1	0.9	-1.4	-2.4	0.6
India	719	931	1,292	1,473	624	710	874	951	2.6	3.2	2.6	0.3	1.1	1.7
Indonesia	669	811	1,084	1,224	726	806	1,211	1,396	2.4	4.6	2.9	-0.1	2.0	0.2
Kazakhstan	52	26	36	44	6,803	2,903	3,371	6,122	-7.2	3.1	4.1	3.7	1.6	2.3
Korea, DPR	1,532	1,294	1,417	1,554	1,065	995	1,231	1,412	-1.6	1.4	1.5	2.6	1.3	2.7
Korea, Republic	3,661	5,252	5,703	5,997	2,299	4,699	7,899	10,879	2.6	-0.2	0.6	1.9	-0.1	1.7
Kyrgyzstan	152	161	177	194	2,786	1,841	2,692	2,978	0.8	0.9	1.7	5.3	0.2	1.9
Lao PDR	428	647	829	1,247	472	618	754	1,058	5.0	4.5	9.5	2.7	-0.9	2.6
Malaysia	1,100	1,403	1,868	1,961	3,894	5,327	8,904	10,745	2.7	3.5	2.0	1.1	1.8	0.7
Micronesia	392	380	370	390	558	526	542	573	-0.9	-0.4	0.8	0.3	-2.3	0.1
Mongolia	7	7	7	9	3,827	3,991	3,471	4,930	0.1	-2.3	5.9	3.1	-2.3	-2.4
Myanmar	596	1,000	1,671	1,656	401	587	1,030	990	5.5	6.6	0.1	2.2	3.9	-5.9
Nepal	704	912	1,241	1,508	463	470	474	518	2.8	2.8	3.9	0.3	1.3	-0.8
New Caledonia (France)	84	91	113	128	646	698	699	764	1.4	0.0	1.1	0.8	-0.2	1.8
Pakistan	595	808	1,059	1,193	1,398	1,587	1,519	1,628	3.5	2.3	3.0	1.1	0.3	1.2
Papua New Guinea	1,908	2,181	2,248	2,431	1,178	1,271	1,268	1,237	2.7	2.0	1.6	0.8	-0.4	0.3
Philippines	1,150	1,454	1,758	1,769	1,175	1,312	1,584	1,596	2.4	2.6	0.7	0.6	1.5	-1.2
Polynesia	704	749	969	1,011	1,283	1,331	1,790	1,983	0.1	1.6	1.2	1.1	2.0	1.2
Solomon Islands	945	1,110	1,070	1,111	706	709	773	732	2.7	3.1	1.0	0.4	0.5	-0.5
Sri Lanka	900	998	1,156	1,142	589	645	757	782	1.1	2.6	0.7	-1.1	1.2	-1.2
Taiwan (China)	5,958	6,651	6,201	6,384	4,984	7,669	9,166	9,163	0.7	-1.2	0.2	1.3	0.4	1.2
Tajikistan	313	173	320	422	1,687	697	1,035	1,295	-5.8	6.6	5.5	0.6	2.1	4.0
Thailand	844	1,265	1,486	1,515	856	1,265	1,688	1,942	3.3	2.2	1.4	1.2	1.1	1.5
Timor-Leste	339	311	396	354	433	441	436	362	-0.3	3.4	-1.9	-1.8	-0.4	-5.5
Turkmenistan	40	54	81	83	2,788	2,117	2,488	2,334	1.1	3.6	0.3	0.0	0.2	0.8
Uzbekistan	259	255	431	568	2,351	2,254	3,684	4,221	-0.5	5.0	5.6	2.0	1.3	2.4
Vanuatu	453	348	418	400	2,296	1,846	2,059	1,920	-1.2	2.5	-0.9	-2.2	0.1	-0.6
Viet Nam	1,590	2,137	2,680	2,890	459	704	960	1,086	5.6	4.3	3.2	-1.0	1.1	1.4

Table 4 continued

Region/country	Land productivity (in constant 2004-2006 US dollars)				Labor productivity (in constant 2004-2006 US dollars)				Output growth (%)			TFP growth (%)		
	1990	2000	2010	2015	1990	2000	2010	2015	1991-2000	2001-2010	2011-2015	1991-2000	2001-2010	2011-2015
MIDDLE EAST AND NORTH AFRICA	**1,073**	**1,344**	**1,596**	**1,738**	**2,565**	**3,491**	**4,484**	**5,240**	**2.7**	**2.5**	**1.8**	**1.3**	**1.5**	**1.2**
Algeria	74	102	194	224	1,438	1,440	2,417	2,731	3.6	6.8	3.0	2.8	3.9	-1.2
Bahrain	2,424	2,831	2,925	4,933	4,849	8,683	6,290	14,141	2.9	-0.3	10.4	4.0	-0.2	8.7
Egypt	4,179	5,235	5,996	6,399	1,719	2,780	3,455	3,953	4.4	2.5	2.1	1.4	0.9	0.5
Iran	217	301	531	580	2,568	3,251	3,750	4,035	3.5	2.7	1.5	2.1	0.7	3.8
Iraq	332	313	349	219	4,833	4,861	6,729	5,162	-1.6	1.0	-6.9	-0.2	0.0	-8.9
Jordan	554	722	1,260	1,410	5,760	6,592	10,610	12,845	2.9	4.9	3.3	1.3	4.0	1.3
Kuwait	643	996	1,639	2,412	10,072	13,402	15,574	21,197	4.9	5.2	7.4	3.3	1.5	4.9
Lebanon	1,762	2,082	1,832	1,737	16,654	30,209	40,422	45,724	1.5	-0.6	-0.5	-0.7	-0.3	-0.4
Libya	53	65	77	81	6,557	9,904	17,704	24,301	2.1	1.6	0.9	2.6	-0.7	8.7
Morocco	167	170	305	336	1,557	1,552	3,066	3,608	0.3	5.6	2.3	-1.1	4.1	0.7
Oman	173	264	280	337	749	1,091	1,287	1,054	5.1	2.8	3.7	4.8	-0.9	0.8
Qatar	448	708	801	1,014	3,902	11,687	6,560	8,317	5.4	1.2	4.7	5.6	-2.3	1.8
Saudi Arabia	20	16	20	19	2,491	4,109	6,599	8,022	1.2	2.3	-0.3	-0.2	2.7	-1.5
Syria	272	408	448	370	3,803	4,888	4,423	3,824	4.2	1.1	-3.8	2.0	-0.4	2.5
Tunisia	282	302	363	483	3,771	3,780	4,465	5,936	1.7	2.3	5.8	-0.7	0.9	4.9
Turkey	677	785	943	1,115	2,600	3,506	4,594	5,734	1.7	1.5	3.1	1.2	2.0	0.9
United Arab Emirates	880	3,222	2,130	1,995	3,196	13,518	4,508	4,587	16.2	-4.4	-2.1	10.7	-7.2	-0.5
West Bank and Gaza	-	1,672	1,873	1,811	-	4,975	4,260	5,343	5.6	-2.9	3.0	-1.9	-0.3	-0.8
Yemen	33	48	79	77	583	606	847	824	3.8	4.9	-0.4	2.2	2.4	1.7

Note: Land productivity is agricultural gross production per hectare of agricultural land; labor productivity is agricultural gross production per economically active person in agriculture.

Projections of Food Production, Consumption, and Hunger

INTERNATIONAL MODEL FOR POLICY ANALYSIS OF AGRICULTURAL COMMODITIES AND TRADE (IMPACT)

Policymakers, analysts, and civil society face increasing challenges to reducing hunger and improving food security in a sustainable way. Modeling alternative future scenarios and assessing their outcomes can help inform policy choices. The International Food Policy Research Institute's IMPACT model is an integrated system of linked economic, climate, water, and crop models that allows for exploration of such scenarios.

METHODOLOGY

At IMPACT's core is a partial equilibrium, multimarket economic model that simulates national and international agricultural markets. Links to climate, water, and crop models support the integrated study of changing environmental, biophysical, and socioeconomic trends, allowing for in-depth analysis of a variety of critical issues of interest to policymakers at national, regional, and global levels. IMPACT benefits from close interactions with scientists at all 15 CGIAR research centers through the Global Futures and Strategic Foresight (GFSF) program, and with other leading global economic modeling efforts around the world through the Agricultural Model Intercomparison and Improvement Project (AgMIP).

Tables 5 and 6 summarize results from the latest IMPACT projections to 2030 and 2050. These results update previous projections by incorporating feedback from the GLOBE general equilibrium model, which captures interactions between the agriculture sector and the wider economy. Results are shown for production, consumption, and trade of major food commodity groups, as well as for the population at risk of hunger, by region and for selected countries. Results are shown for two "baseline" scenarios—one considers the impacts of climate change, while the other assumes no climate change (for comparison). Results for additional countries can be found at the IMPACT website.

KEY FINDINGS FROM THE LATEST IMPACT PROJECTIONS

The baseline projections from IMPACT indicate that global food production will grow by about 59 percent over 2010 levels by 2050 in the context of climate change—8 percentage points less than would be the case without climate change (Table 5). Production will grow more rapidly in developing countries, particularly in Africa, than in developed countries. Even with population growth and climate change, per capita consumption (assumed equivalent to per capita availability) is projected to increase by about 10 percent globally to more than 3,000 kilocalories per day. But differences in access to food within and between countries mean that nearly 500 million people will remain at risk of hunger. In Africa south of the Sahara, an additional 41 million people are projected to be at risk of hunger in 2050 as a result of climate change—27 percent more than would be at risk in the absence of climate change.

Despite the impacts of climate change, meat production is projected to grow by 65 percent globally by 2050, and by 77 percent in developing countries. Per capita consumption levels in developing countries, however, will remain under half of those in developed countries (Table 6). Production of fruits and vegetables, pulses, and oilseeds will grow even more rapidly, by about 80 percent globally and more than doubling in some regions. Per capita consumption of fruits and vegetables in developing countries is projected to surpass that of developed countries by 2050, with important benefits for nutrition and health. Production of cereals and roots and tubers will grow more slowly, by around 40 percent globally but roughly doubling in Africa south of the Sahara. Developing countries as a group will become larger net importers of food from developed countries.

In addition to the indicators presented here, IMPACT also explores changes in prices, land and water use, greenhouse gas emissions, and other socioeconomic and environmental indicators. For example, when the impacts of climate change are considered, prices are projected to rise by about 50 percent for most food commodity groups by 2050—about double the increase projected in the absence of climate change.

MORE INFORMATION

More information on these results, and on the results of alternative scenarios exploring different population, income, policy, investment, and technology pathways, can be found online. Results for all 158 countries and regions modeled are available as well as information on IMPACT, the GFSF program, and recent publications.

VISIT ONLINE
IFPRI IMPACT:
https://www.ifpri.org/program/impact-model

Global Futures and Strategic Foresight:
http://globalfutures.cgiar.org/

IMPACT documentation:
http://www.ifpri.org/publication/international-model-policy-analysis-agricultural-commodities-and-trade-impact-model-0

DOWNLOAD DATA
https://dataverse.harvard.edu/dataverse/impact

CONTACT
IMPACT (IFPRI-Impact-Model@cgiar.org)

TABLE 5 IMPACT projections of food production, consumption, and hunger to 2050, with and without climate change

	Aggregate food production (index, 2010 = 1.00)					Per capita food consumption (KCAL per capita per day)					Hunger (millions of people at risk)				
		Without climate change		With climate change			Without climate change		With climate change			Without climate change		With climate change	
	2010	2030	2050	2030	2050	2010	2030	2050	2030	2050	2010	2030	2050	2030	2050
WORLD	**1.00**	**1.37**	**1.67**	**1.33**	**1.59**	**2,795**	**3,032**	**3,191**	**2,979**	**3,070**	**838.1**	**528.2**	**405.8**	**598.0**	**482.6**
Developing countries	1.00	1.42	1.77	1.40	1.71	2,683	2,961	3,137	2,904	3,008	823.3	513.3	392.2	582.4	466.9
Developed countries	1.00	1.22	1.42	1.15	1.27	3,384	3,439	3,513	3,407	3,441	14.8	14.9	13.6	15.6	15.7
ASIA AND THE PACIFIC	**1.00**	**1.39**	**1.68**	**1.39**	**1.66**	**2,656**	**3,003**	**3,185**	**2,948**	**3,057**	**539.8**	**249.8**	**181.8**	**285.0**	**207.4**
East Asia	**1.00**	**1.26**	**1.39**	**1.28**	**1.43**	**3,009**	**3,509**	**3,628**	**3,453**	**3,507**	**187.2**	**59.2**	**54.7**	**60.3**	**56.8**
China	1.00	1.26	1.38	1.28	1.43	3,044	3,604	3,733	3,545	3,606	173.4	44.8	41.0	44.7	41.0
Japan	1.00	1.20	1.41	1.25	1.53	2,770	2,787	2,842	2,757	2,776	2.3	2.0	1.2	2.3	1.9
Korea, Rep.	1.00	1.26	1.46	1.27	1.45	3,139	3,347	3,429	3,310	3,348	0.6	0.4	0.4	0.4	0.4
South Asia	**1.00**	**1.60**	**2.10**	**1.54**	**1.97**	**2,361**	**2,669**	**2,959**	**2,617**	**2,826**	**268.5**	**138.3**	**87.7**	**165.2**	**99.0**
Afghanistan	1.00	1.37	1.85	1.38	1.87	2,149	2,239	2,452	2,199	2,332	7.0	9.4	7.9	10.3	10.9
Bangladesh	1.00	1.44	1.70	1.36	1.53	2,426	2,714	2,911	2,652	2,776	26.0	11.3	6.9	14.9	8.9
India	1.00	1.68	2.25	1.62	2.11	2,354	2,697	2,998	2,644	2,860	189.7	73.9	45.0	93.1	44.9
Nepal	1.00	1.36	1.69	1.39	1.77	2,425	2,695	3,186	2,613	2,984	2.7	2.0	0.8	2.4	1.6
Pakistan	1.00	1.24	1.45	1.20	1.35	2,379	2,540	2,862	2,507	2,766	37.6	38.0	24.4	40.5	29.2
Southeast Asia and Pacific	**1.00**	**1.47**	**1.87**	**1.44**	**1.81**	**2,551**	**2,852**	**3,051**	**2,791**	**2,920**	**84.1**	**52.3**	**39.4**	**59.5**	**51.7**
Indonesia	1.00	1.62	2.02	1.63	2.03	2,540	2,990	3,281	2,902	3,090	32.4	12.9	7.2	15.6	11.4
Malaysia	1.00	1.83	2.94	1.78	2.81	2,838	3,173	3,462	3,141	3,380	0.9	0.8	0.9	0.8	0.9
Philippines	1.00	1.33	1.68	1.31	1.65	2,503	2,641	2,777	2,598	2,682	12.1	12.2	11.0	13.4	13.3
Thailand	1.00	1.18	1.27	1.12	1.14	2,742	3,012	3,183	2,972	3,099	6.2	3.1	1.8	3.5	2.3
Viet Nam	1.00	1.25	1.36	1.20	1.24	2,512	2,710	2,828	2,649	2,699	12.9	9.5	7.2	11.0	9.9

Table 5 continued

	Aggregate food production (index, 2010 = 1.00)					Per capita food consumption (KCAL per capita per day)					Hunger (millions of people at risk)				
		Without climate change		With climate change			Without climate change		With climate change			Without climate change		With climate change	
	2010	2030	2050	2030	2050	2010	2030	2050	2030	2050	2010	2030	2050	2030	2050
AFRICA AND MIDDLE EAST	**1.00**	**1.60**	**2.23**	**1.55**	**2.10**	**2,623**	**2,795**	**3,002**	**2,731**	**2,866**	**238.7**	**229.8**	**185.0**	**260.2**	**229.9**
Africa south of the Sahara	**1.00**	**1.65**	**2.36**	**1.57**	**2.16**	**2,358**	**2,587**	**2,853**	**2,513**	**2,703**	**209.5**	**195.7**	**150.5**	**224.4**	**191.5**
Congo, Dem. Rep.	1.00	1.72	2.49	1.67	2.38	1,943	2,392	2,998	2,326	2,852	37.6	20.3	6.6	25.1	6.6
Ethiopia	1.00	1.66	2.46	1.66	2.47	2,066	2,307	2,614	2,261	2,519	32.7	32.3	22.5	35.0	27.3
Kenya	1.00	1.69	2.84	1.72	2.84	2,133	2,395	2,708	2,293	2,504	10.2	8.9	5.0	11.0	8.7
Nigeria	1.00	1.62	2.31	1.56	2.15	2,751	2,943	3,136	2,857	2,964	9.7	8.5	11.6	10.9	11.8
South Africa	1.00	1.49	1.84	1.48	1.78	2,962	3,229	3,397	3,155	3,256	1.9	1.5	1.6	1.5	1.6
Sudan	1.00	1.68	2.39	1.48	1.91	2,329	2,465	2,714	2,430	2,635	11.4	12.7	9.0	13.7	10.9
Tanzania, United Rep.	1.00	1.63	2.39	1.55	2.19	2,178	2,396	2,602	2,305	2,432	15.6	17.8	17.8	20.6	23.3
Uganda	1.00	1.88	3.01	1.76	2.66	2,391	2,585	2,796	2,519	2,664	8.5	10.4	11.3	11.8	13.9
Middle East and North Africa	**1.00**	**1.52**	**2.03**	**1.51**	**2.01**	**3,125**	**3,250**	**3,377**	**3,207**	**3,275**	**29.3**	**34.2**	**34.5**	**35.8**	**38.4**
Algeria	1.00	1.55	2.05	1.44	1.77	2,977	3,098	3,163	3,062	3,075	1.9	1.9	1.9	2.0	2.2
Egypt	1.00	1.49	2.00	1.44	1.92	3,395	3,580	3,783	3,518	3,641	1.6	2.2	2.5	2.2	2.5
Iran	1.00	1.47	1.92	1.50	1.99	3,079	3,109	3,228	3,068	3,129	4.7	5.2	4.4	5.7	5.3
Iraq	1.00	1.80	3.19	1.77	3.11	2,342	2,651	2,773	2,616	2,681	7.8	7.5	8.5	7.9	9.6
Morocco	1.00	1.62	2.29	1.44	1.86	3,287	3,592	3,856	3,548	3,744	1.7	1.9	2.0	1.9	2.0
Saudi Arabia	1.00	1.74	2.68	1.73	2.63	2,936	3,055	3,128	3,022	3,054	1.3	1.4	1.5	1.5	1.8
Turkey	1.00	1.44	1.68	1.46	1.75	3,596	3,661	3,698	3,620	3,599	1.8	2.2	2.4	2.2	2.4
THE AMERICAS	**1.00**	**1.35**	**1.65**	**1.26**	**1.46**	**3,188**	**3,290**	**3,392**	**3,244**	**3,299**	**42.5**	**35.7**	**27.7**	**39.4**	**32.8**
Latin America and the Caribbean	**1.00**	**1.44**	**1.79**	**1.40**	**1.68**	**2,878**	**3,036**	**3,184**	**2,984**	**3,080**	**39.5**	**32.1**	**24.0**	**35.9**	**28.8**
Argentina	1.00	1.42	1.73	1.42	1.72	3,171	3,327	3,426	3,297	3,355	0.7	0.6	0.7	0.6	0.7
Brazil	1.00	1.51	1.92	1.41	1.65	3,142	3,336	3,492	3,292	3,399	3.7	3.2	3.1	3.2	3.1
Colombia	1.00	1.37	1.62	1.43	1.76	2,645	2,804	2,957	2,758	2,867	5.0	3.9	2.7	4.5	3.6
Mexico	1.00	1.33	1.56	1.29	1.48	3,040	3,134	3,240	3,055	3,099	5.3	5.4	5.3	6.1	6.1
Peru	1.00	1.47	1.79	1.69	2.38	2,472	2,752	2,886	2,697	2,776	3.6	2.0	1.4	2.3	1.9
Venezuela	1.00	1.39	1.72	1.29	1.49	2,536	2,626	2,763	2,578	2,669	1.4	1.3	0.7	1.6	1.2
North America	**1.00**	**1.28**	**1.54**	**1.16**	**1.28**	**3,714**	**3,725**	**3,735**	**3,690**	**3,660**	**3.0**	**3.6**	**3.7**	**3.6**	**4.0**
EUROPE & FORMER SOVIET UNION	**1.00**	**1.16**	**1.26**	**1.13**	**1.21**	**3,275**	**3,390**	**3,491**	**3,360**	**3,417**	**17.1**	**13.0**	**11.4**	**13.4**	**12.4**
Former Soviet Union	**1.00**	**1.21**	**1.32**	**1.17**	**1.27**	**3,092**	**3,321**	**3,423**	**3,287**	**3,339**	**9.7**	**5.9**	**5.2**	**6.2**	**5.5**
Russia	1.00	1.20	1.32	1.18	1.31	3,227	3,450	3,532	3,417	3,454	1.8	1.2	1.1	1.2	1.2
Ukraine	1.00	1.16	1.23	1.08	1.11	3,201	3,434	3,581	3,399	3,499	0.6	0.4	0.3	0.4	0.3
Uzbekistan	1.00	1.30	1.49	1.28	1.46	2,563	2,849	3,024	2,819	2,934	2.4	0.8	0.8	0.9	0.8
Europe	**1.00**	**1.14**	**1.24**	**1.11**	**1.18**	**3,370**	**3,424**	**3,523**	**3,396**	**3,454**	**7.4**	**7.0**	**6.2**	**7.3**	**6.9**

Note: World and regional figures include other regions and countries not reported separately. Country-level details are available online. Aggregate food production is an index, by weight, of cereals, meats, fruits and vegetables, oilseeds, pulses, and roots and tubers (which are reported separately in Table 6). Per capita food consumption is a projection of daily dietary energy supply. Estimates of the number of people at risk of hunger are based on a quadratic specification of the relationship between national-level calorie supply and the share of population that is undernourished as defined by the FAO. Values reported for 2010 are calibrated model results. Projections for 2030 and 2050 assume changes in population and income as reflected in the IPCC's Shared Socioeconomic Pathway 2. Climate change impacts are simulated using the IPCC's Representative Concentration Pathway 8.5 and the HadGEM general circulation model. Further documentation is available at www.ifpri.org/program/impact-model.

TABLE 6 IMPACT projections of production, consumption, and net trade to 2050 by commodity group, with and without climate change

	Total production (million metric tons)					Per capita food consumption (kg per capita per year)					Net trade (million metric tons)				
		Without climate change		With climate change			Without climate change		With climate change			Without climate change		With climate change	
	2010	2030	2050	2030	2050	2010	2030	2050	2030	2050	2010	2030	2050	2030	2050
CEREALS															
WORLD	2,155	2,746	3,235	2,622	2,986	143.5	146.7	148.3	143.4	140.4	0.0	0.0	0.0	0.0	0.0
Developing countries	1,390	1,826	2,154	1,802	2,106	148.7	151.6	153.0	148.0	144.5	-86.6	-124.0	-224.3	-60.8	-93.0
Developed countries	765	920	1,081	820	880	116.3	118.3	120.4	116.7	115.6	86.6	124.0	224.3	60.8	93.0
ASIA AND PACIFIC	859	1,067	1,195	1,048	1,164	148.7	152.1	154.3	148.9	146.1	-39.7	-69.7	-129.4	-28.7	-8.7
East Asia	393	451	479	464	511	145.2	148.2	147.3	145.4	140.3	-43.3	-63.3	-74.5	-6.9	61.9
South Asia	279	384	454	362	414	148.5	150.7	154.1	147.5	145.9	-5.1	-8.3	-52.7	-21.7	-65.9
Southeast Asia and Pacific	187	232	262	221	239	158.1	164.6	167.6	159.7	157.2	8.6	1.9	-2.2	0.0	-4.6
AFRICA AND MIDDLE EAST	229	337	428	328	408	149.3	151.0	151.5	146.6	142.4	-91.5	-157.6	-261.3	-152.8	-239.1
Africa south of the Sahara	114	184	254	179	239	121.8	129.3	134.4	124.0	124.3	-32.2	-63.5	-119.9	-57.6	-102.4
West	49	79	110	75	99	143.5	152.4	155.3	146.6	144.5	-13.7	-29.8	-60.3	-28.9	-56.5
Central	7	12	18	12	16	59.3	65.4	68.9	62.3	63.0	-3.1	-6.3	-11.8	-5.9	-10.5
East	39	65	91	64	91	115.7	125.6	134.1	119.5	122.8	-8.7	-17.1	-31.9	-13.6	-21.6
Southern	13	18	21	19	23	182.8	194.8	201.5	187.3	187.3	-3.5	-7.1	-12.5	-4.6	-7.2
Middle East and North Africa	114	153	174	149	169	201.4	198.3	194.4	195.9	187.6	-59.3	-94.1	-141.4	-95.1	-136.7
North Africa	42	55	62	49	50	204.7	202.5	198.7	199.6	191.3	-30.6	-46.4	-68.5	-49.9	-73.0
Middle East	78	107	126	108	129	183.2	179.1	175.8	177.1	169.8	-31.8	-50.9	-76.1	-49.7	-70.1
THE AMERICAS	600	817	1,033	714	806	120.6	121.7	121.5	118.9	115.3	100.8	189.9	312.3	132.5	131.5
Latin America and the Caribbean	164	245	322	236	295	128.0	129.6	129.8	126.0	122.8	-23.4	-18.4	-5.8	-17.6	-57.1
Caribbean	2	2	3	2	3	103.7	104.9	105.3	102.4	100.1	-5.6	-7.3	-9.1	-6.9	-8.0
Central America	38	52	66	51	64	156.9	156.6	155.2	150.2	144.0	-23.4	-28.3	-29.0	-25.0	-25.3
South America	125	191	254	182	227	118.7	120.5	120.9	118.1	115.6	5.6	17.2	32.3	14.3	-23.9
North America	436	572	711	478	511	108.2	108.3	107.8	106.6	102.9	124.3	208.3	318.1	150.0	188.7
EUROPE AND FORMER SOVIET UNION	467	525	579	532	609	135.9	140.6	144.2	139.1	139.1	30.4	37.4	78.4	49.0	116.2
Former Soviet Union	156	206	244	217	271	162.1	170.8	174.5	169.3	168.9	21.5	62.3	101.5	77.2	136.7
Europe	311	319	334	315	338	122.3	125.5	129.6	124.1	124.8	8.9	-24.9	-23.1	-28.2	-20.5
MEATS															
WORLD	274	381	460	379	453	39.4	45.6	49.5	45.3	48.8	0.0	0.0	0.0	0.0	0.0
Developing countries	174	254	312	252	307	30.5	37.7	41.9	37.4	41.1	-3.6	-14.4	-21.5	-14.0	-19.6
Developed countries	100	127	148	127	145	86.5	91.1	95.8	90.8	95.2	3.6	14.4	21.5	14.0	19.6
ASIA AND PACIFIC	109	150	166	149	164	30.3	39.6	43.3	39.3	42.4	-7.0	-25.3	-34.7	-25.1	-33.3
East Asia	79	99	93	98	91	56.5	76.3	81.3	75.7	80.2	-9.2	-22.5	-25.6	-22.8	-26.0
South Asia	10	19	31	19	31	6.0	10.7	17.8	10.5	17.0	0.2	-2.9	-11.4	-2.6	-9.9
Southeast Asia and Pacific	20	32	43	32	42	28.8	41.6	49.6	41.3	48.8	2.0	0.1	2.3	0.3	2.6
AFRICA AND MIDDLE EAST	22	40	66	40	65	18.3	23.7	31.3	23.5	30.7	-2.7	-6.1	-12.9	-6.0	-12.1
Africa south of the Sahara	11	20	35	20	34	13.0	18.1	26.8	17.9	26.2	-0.4	-3.6	-13.5	-3.4	-12.6
West	3	6	11	6	11	10.2	16.2	26.6	15.9	25.5	-0.3	-1.9	-7.3	-1.8	-6.6

Table 6 continued

	Total production (million metric tons)					Per capita food consumption (kg per capita per year)					Net trade (million metric tons)				
		Without climate change		With climate change			Without climate change		With climate change			Without climate change		With climate change	
	2010	2030	2050	2030	2050	2010	2030	2050	2030	2050	2010	2030	2050	2030	2050
Central	1	1	2	1	2	9.1	12.2	17.0	12.1	16.7	-0.4	-1.0	-2.1	-1.0	-2.0
East	3	6	10	6	10	10.3	14.4	22.5	14.3	22.1	0.0	-1.1	-4.9	-1.1	-4.7
Southern	2	4	5	4	5	45.2	61.0	73.3	60.6	72.4	-0.2	-0.1	-0.1	-0.1	-0.1
Middle East and North Africa	11	20	31	19	31	28.3	36.0	42.4	35.8	41.9	-2.3	-2.5	0.7	-2.5	0.4
North Africa	5	10	17	10	16	22.6	32.0	42.9	31.8	42.4	0.0	0.3	2.0	0.3	2.0
Middle East	7	13	20	13	20	31.2	38.0	43.5	37.8	43.0	-1.9	-2.2	-0.5	-2.2	-0.7
THE AMERICAS	**89**	**127**	**158**	**126**	**155**	**82.2**	**88.0**	**93.0**	**87.6**	**92.3**	**11.5**	**29.1**	**44.5**	**28.6**	**42.6**
Latin America and the Caribbean	44	67	85	66	83	61.4	69.9	76.6	69.4	75.7	7.2	16.7	25.9	16.5	25.3
Caribbean	1	2	2	2	2	34.5	43.3	52.4	43.0	51.7	-0.1	-0.2	0.0	-0.1	0.0
Central America	7	10	13	10	13	51.0	58.4	65.8	58.0	65.0	-0.9	-0.9	-0.8	-0.8	-0.7
South America	36	55	69	54	68	68.1	76.9	83.2	76.5	82.3	8.2	17.7	26.7	17.5	26.0
North America	45	61	73	60	72	117.6	119.0	120.2	118.7	119.6	4.4	12.5	18.7	12.1	17.3
EUROPE AND FORMER SOVIET UNION	**54**	**64**	**69**	**64**	**69**	**67.5**	**72.0**	**76.3**	**71.6**	**75.7**	**-1.8**	**2.3**	**3.0**	**2.5**	**2.8**
Former Soviet Union	10	12	14	12	13	46.0	55.3	59.5	55.0	59.0	-3.0	-3.3	-3.2	-3.3	-3.2
Europe	44	52	56	52	55	78.6	80.3	84.4	79.9	83.7	1.2	5.6	6.2	5.7	6.1
FRUITS AND VEGETABLES															
WORLD	**1,592**	**2,334**	**3,044**	**2,290**	**2,913**	**196.2**	**240.0**	**284.7**	**235.5**	**272.2**	**0.0**	**0.0**	**0.0**	**0.0**	**0.0**
Developing countries	1,304	1,952	2,554	1,918	2,447	191.2	239.7	290.6	235.0	277.2	15.1	-20.1	-90.5	-14.4	-76.9
Developed countries	288	383	490	372	466	222.8	241.4	248.6	238.0	242.0	-15.1	20.1	90.5	14.4	76.9
ASIA AND PACIFIC	**868**	**1,259**	**1,586**	**1,258**	**1,562**	**209.7**	**278.7**	**358.7**	**273.1**	**341.1**	**-44.8**	**-141.4**	**-279.9**	**-114.4**	**-213.0**
East Asia	609	800	938	820	978	351.0	432.7	430.1	426.4	419.0	-20.4	-5.4	192.7	26.5	252.5
South Asia	158	318	467	300	411	104.7	197.7	366.8	192.2	342.6	-29.7	-127.5	-466.1	-132.9	-460.6
Southeast Asia and Pacific	101	141	181	138	173	134.0	176.0	205.1	171.9	195.3	5.3	-8.5	-6.6	-8.0	-4.9
AFRICA AND MIDDLE EAST	**251**	**436**	**661**	**421**	**617**	**155.9**	**171.4**	**190.1**	**168.2**	**182.7**	**1.9**	**33.1**	**77.8**	**26.1**	**57.2**
Africa south of the Sahara	101	188	301	173	258	95.4	119.7	150.0	116.7	142.7	-1.0	-9.3	-34.1	-18.9	-59.7
West	40	74	118	70	105	117.2	145.3	174.4	141.6	165.7	0.3	-3.5	-14.8	-5.7	-21.3
Central	10	17	27	16	22	66.0	82.4	103.1	80.1	97.5	0.1	-1.3	-4.4	-2.5	-7.6
East	36	70	121	65	106	82.2	105.5	138.5	102.9	131.6	-1.2	-5.4	-12.9	-8.2	-20.6
Southern	9	15	21	14	17	76.2	89.2	98.3	87.3	94.2	2.9	6.4	10.1	5.2	7.1
Middle East and North Africa	150	248	361	248	359	270.2	284.3	290.5	280.5	283.0	3.0	42.4	111.9	45.1	116.9
North Africa	57	99	149	89	124	228.9	250.1	270.3	246.6	262.8	0.0	16.5	44.1	8.1	22.8
Middle East	98	160	226	167	243	278.0	287.3	291.0	283.5	283.5	-0.2	20.3	55.6	29.2	76.8
THE AMERICAS	**255**	**351**	**447**	**338**	**419**	**187.0**	**212.1**	**226.7**	**208.4**	**219.0**	**49.2**	**74.6**	**123.8**	**66.6**	**107.1**
Latin America and the Caribbean	164	236	299	224	271	159.6	182.9	202.9	179.6	195.5	46.3	76.3	108.3	66.9	86.8
Caribbean	12	15	18	14	15	192.6	218.7	245.0	216.1	239.1	2.3	3.4	4.6	1.9	1.8
Central America	46	59	67	55	58	165.5	180.0	196.8	176.7	189.6	15.4	18.6	18.1	15.1	10.9
South America	107	162	214	156	198	154.2	181.0	202.0	177.6	194.5	28.7	54.3	85.7	49.9	74.0

Table 6 continued

	Total production (million metric tons)					Per capita food consumption (kg per capita per year)					Net trade (million metric tons)				
		Without climate change		With climate change			Without climate change		With climate change			Without climate change		With climate change	
	2010	2030	2050	2030	2050	2010	2030	2050	2030	2050	2010	2030	2050	2030	2050
North America	91	114	147	114	148	233.6	262.4	265.9	257.9	257.7	2.9	-1.7	15.4	-0.3	20.3
EUROPE AND FORMER SOVIET UNION	**218**	**289**	**351**	**274**	**314**	**209.2**	**230.9**	**241.8**	**227.8**	**235.7**	**-6.3**	**33.7**	**78.4**	**21.7**	**48.7**
Former Soviet Union	62	81	95	79	89	181.6	223.0	239.5	219.7	232.5	0.1	5.0	14.0	3.8	10.4
Europe	156	208	255	195	225	223.5	234.8	242.9	231.8	237.2	-6.4	28.7	64.4	17.9	38.2
OILSEEDS															
WORLD	**673**	**1,033**	**1,293**	**1,014**	**1,246**	**6.8**	**8.2**	**7.8**	**7.9**	**7.3**	**0.0**	**0.0**	**0.0**	**0.0**	**0.0**
Developing countries	525	842	1,079	830	1,047	7.0	8.6	8.2	8.3	7.6	-3.0	-8.5	-11.5	-7.2	-9.1
Developed countries	148	191	214	184	199	5.6	5.9	5.9	5.6	5.5	3.0	8.5	11.5	7.2	9.1
ASIA AND PACIFIC	**322**	**536**	**713**	**532**	**700**	**8.1**	**10.4**	**9.5**	**10.0**	**8.9**	**-35.4**	**-59.6**	**-69.9**	**-55.7**	**-61.5**
East Asia	49	63	68	64	70	10.9	15.9	15.1	15.4	14.3	-44.3	-62.8	-66.7	-59.1	-59.2
South Asia	41	52	57	50	52	3.6	4.5	4.3	4.3	4.0	0.5	-4.5	-9.7	-4.6	-9.6
Southeast Asia and Pacific	231	421	589	417	578	13.1	14.7	14.6	14.3	13.9	8.4	7.7	6.4	8.0	7.3
AFRICA AND MIDDLE EAST	**61**	**101**	**126**	**98**	**118**	**5.5**	**6.4**	**7.2**	**6.1**	**6.5**	**-6.1**	**-8.8**	**-13.5**	**-8.1**	**-11.5**
Africa south of the Sahara	53	90	113	86	104	5.9	6.8	7.7	6.5	6.9	0.2	-1.2	-4.6	-1.0	-3.8
West	43	74	94	72	87	8.1	9.3	10.1	8.8	9.1	0.3	-0.5	-2.7	-0.4	-2.4
Central	4	6	8	6	7	9.0	10.0	10.6	9.4	9.5	0.1	0.1	0.1	0.2	0.4
East	4	6	7	6	7	3.7	4.4	5.3	4.2	4.8	0.1	-0.3	-1.3	-0.2	-0.9
Southern	1	1	2	1	1	1.9	2.1	2.1	2.0	2.0	-0.2	-0.3	-0.3	-0.2	-0.2
Middle East and North Africa	9	12	14	12	14	4.7	5.5	6.0	5.3	5.5	-6.3	-7.6	-8.8	-7.0	-7.6
North Africa	4	6	7	5	6	4.6	5.3	5.7	5.1	5.2	-1.5	-1.8	-2.2	-1.7	-2.1
Middle East	5	7	9	8	9	4.6	5.2	5.6	5.0	5.2	-5.0	-6.1	-7.2	-5.7	-6.2
THE AMERICAS	**235**	**323**	**371**	**313**	**348**	**6.8**	**6.7**	**6.5**	**6.5**	**6.0**	**58.7**	**83.7**	**97.5**	**77.8**	**84.7**
Latin America and the Caribbean	126	184	215	179	205	6.6	6.4	6.0	6.2	5.5	27.2	46.3	56.6	43.4	49.4
Caribbean	1	1	1	1	1	7.8	8.0	7.8	7.7	7.4	-0.3	-0.3	-0.4	-0.3	-0.4
Central America	6	9	10	8	9	3.2	3.1	3.1	3.0	2.9	-5.5	-6.0	-6.5	-5.7	-5.9
South America	119	174	204	170	195	7.9	7.7	7.1	7.3	6.5	33.0	52.6	63.5	49.4	55.6
North America	110	139	155	133	144	7.1	7.2	7.2	7.0	6.7	31.5	37.3	40.8	34.4	35.3
EUROPE AND FORMER SOVIET UNION	**55**	**72**	**83**	**71**	**80**	**2.6**	**2.7**	**2.9**	**2.6**	**2.7**	**-17.2**	**-15.3**	**-14.1**	**-14.0**	**-11.7**
Former Soviet Union	14	19	22	19	23	1.1	1.2	1.2	1.2	1.2	-0.4	0.5	1.4	0.7	1.8
Europe	40	53	60	52	58	3.3	3.5	3.6	3.3	3.4	-16.8	-15.8	-15.4	-14.8	-13.5
PULSES															
WORLD	**66**	**94**	**121**	**92**	**117**	**6.2**	**7.5**	**8.9**	**7.5**	**8.8**	**0.0**	**0.0**	**0.0**	**0.0**	**0.0**
Developing countries	52	74	97	72	91	6.7	8.2	9.8	8.1	9.6	-2.8	-6.5	-9.9	-7.6	-12.7
Developed countries	14	19	24	20	26	3.5	3.8	4.0	3.8	4.0	2.8	6.5	9.9	7.6	12.7
ASIA AND PACIFIC	**28**	**37**	**44**	**36**	**42**	**5.2**	**6.2**	**7.3**	**6.2**	**7.2**	**-0.5**	**-3.3**	**-5.2**	**-3.1**	**-5.2**
East Asia	6	8	11	8	12	1.5	1.9	2.1	1.9	2.1	0.5	1.8	4.7	2.2	5.6
South Asia	16	21	24	20	23	9.4	10.6	11.7	10.5	11.4	-2.9	-6.1	-10.1	-6.2	-10.5

Table 6 continued

	Total production (million metric tons)					Per capita food consumption (kg per capita per year)					Net trade (million metric tons)				
		Without climate change		With climate change			Without climate change		With climate change			Without climate change		With climate change	
	2010	2030	2050	2030	2050	2010	2030	2050	2030	2050	2010	2030	2050	2030	2050
Southeast Asia and Pacific	7	8	8	8	8	3.1	3.6	3.9	3.5	3.8	1.9	1.0	0.2	0.8	-0.3
AFRICA AND MIDDLE EAST	**16**	**25**	**35**	**23**	**32**	**9.7**	**11.3**	**13.4**	**11.2**	**13.1**	**-1.9**	**-5.6**	**-11.6**	**-6.4**	**-13.4**
Africa south of the Sahara	12	19	28	18	27	10.4	12.3	14.7	12.1	14.3	-0.9	-4.0	-9.4	-3.9	-9.2
West	5	9	16	9	14	8.5	9.8	11.6	9.6	11.0	0.3	0.1	-0.3	0.0	-0.6
Central	1	2	2	2	2	6.7	7.4	8.7	7.3	8.4	-0.1	-0.2	-0.3	-0.2	-0.2
East	5	7	9	7	10	15.3	18.2	22.0	18.0	21.5	-0.7	-3.3	-7.9	-3.1	-7.4
Southern	0	0	0	0	0	3.8	4.2	4.5	4.1	4.4	-0.1	-0.1	0.0	-0.1	0.0
Middle East and North Africa	4	6	7	5	5	8.2	9.2	10.0	9.2	10.0	-1.0	-1.7	-2.3	-2.5	-4.2
North Africa	1	2	2	1	2	8.2	9.7	11.4	9.8	11.4	-1.1	-1.8	-2.6	-2.1	-3.2
Middle East	3	4	5	4	4	8.3	8.8	9.2	8.8	9.2	-0.2	-0.4	-0.6	-1.0	-2.0
THE AMERICAS	**14**	**21**	**28**	**21**	**30**	**8.9**	**9.7**	**10.4**	**9.7**	**10.3**	**3.2**	**7.2**	**12.5**	**8.0**	**14.4**
Latin America and the Caribbean	7	11	16	10	15	11.4	12.6	13.8	12.5	13.6	-0.7	1.1	4.4	0.7	3.0
Caribbean	0	0	1	0	1	12.4	13.4	14.8	13.3	14.6	-0.2	-0.2	0.0	-0.2	-0.1
Central America	2	3	4	3	3	12.5	13.6	15.2	13.5	15.0	-0.4	-0.2	0.0	-0.4	-0.4
South America	5	8	12	7	11	10.8	12.1	13.1	12.0	12.9	0.0	1.5	4.4	1.2	3.6
North America	7	10	12	11	15	4.6	4.8	4.9	4.8	4.9	3.8	6.0	8.0	7.3	11.3
EUROPE AND FORMER SOVIET UNION	**8**	**11**	**14**	**11**	**14**	**2.6**	**2.7**	**2.8**	**2.7**	**2.8**	**-0.8**	**1.7**	**4.4**	**1.6**	**4.2**
Former Soviet Union	3	4	5	4	5	1.5	1.6	1.6	1.6	1.6	0.4	1.3	2.4	1.3	2.5
Europe	5	7	9	7	8	3.1	3.3	3.4	3.3	3.4	-1.2	0.5	2.0	0.3	1.7
ROOTS AND TUBERS															
WORLD	**780**	**1,006**	**1,185**	**962**	**1,101**	**65.0**	**70.5**	**73.4**	**67.7**	**68.9**	**0.0**	**0.0**	**0.0**	**0.0**	**0.0**
Developing countries	682	897	1,068	857	995	65.8	72.4	75.7	69.5	71.1	5.6	-0.6	-5.4	-0.6	-0.6
Developed countries	97	109	118	105	106	61.2	59.8	59.3	57.5	56.1	-5.6	0.6	5.4	0.6	0.6
ASIA AND PACIFIC	**298**	**351**	**365**	**356**	**379**	**46.9**	**50.9**	**49.5**	**48.3**	**45.6**	**-4.9**	**-23.4**	**-18.8**	**1.4**	**28.7**
East Asia	181	201	185	201	182	71.4	76.3	73.5	72.9	68.7	-18.5	-14.2	-0.1	-3.4	11.6
South Asia	50	75	103	79	119	27.3	35.7	38.0	33.0	33.8	-6.2	-24.2	-30.6	-12.1	2.2
Southeast Asia and Pacific	67	76	77	76	78	37.5	39.4	39.9	38.6	38.6	19.9	15.0	11.9	16.9	14.9
AFRICA AND MIDDLE EAST	**245**	**377**	**524**	**361**	**485**	**109.3**	**117.3**	**123.1**	**113.8**	**116.9**	**-1.8**	**-13.0**	**-31.6**	**-16.5**	**-39.4**
Africa south of the Sahara	224	349	490	332	449	146.4	152.7	156.1	149.0	149.1	-1.1	-11.0	-29.0	-17.7	-42.8
West	133	207	297	201	280	197.5	199.0	198.8	194.7	190.8	1.5	-4.3	-11.7	-4.1	-9.8
Central	37	59	80	56	72	172.5	170.6	166.7	167.2	160.2	1.0	2.6	-2.2	0.1	-8.2
East	50	78	107	71	91	129.6	138.5	142.0	134.5	134.4	-3.2	-9.4	-15.3	-13.9	-24.6
Southern	3	4	5	4	5	36.8	37.7	38.7	36.5	37.0	0.0	0.7	1.3	0.9	1.3
Middle East and North Africa	21	28	34	29	36	39.0	39.9	40.5	37.0	36.3	-0.8	-2.1	-2.6	1.2	3.4
North Africa	9	14	18	15	20	33.7	38.3	42.1	35.6	37.8	-0.1	0.2	-0.3	2.0	4.0
Middle East	13	15	18	15	17	38.8	36.6	35.2	34.0	31.7	-1.1	-2.8	-3.4	-1.6	-2.1
THE AMERICAS	**86**	**112**	**130**	**110**	**127**	**55.7**	**54.5**	**53.0**	**52.4**	**50.0**	**-0.3**	**10.2**	**19.1**	**14.1**	**26.1**

Table 6 continued

	Total production (million metric tons)					Per capita food consumption (kg per capita per year)					Net trade (million metric tons)				
		Without climate change		With climate change			Without climate change		With climate change			Without climate change		With climate change	
	2010	2030	2050	2030	2050	2010	2030	2050	2030	2050	2010	2030	2050	2030	2050
Latin America and the Caribbean	60	82	97	83	98	51.1	49.9	47.9	48.3	45.7	0.2	11.5	20.4	16.1	29.5
Caribbean	3	5	7	5	6	61.4	59.0	56.6	58.2	55.2	0.1	1.7	3.5	1.3	2.6
Central America	3	5	7	5	6	17.4	18.5	19.9	17.5	18.2	-0.2	0.2	0.8	0.2	0.3
South America	53	72	83	74	87	63.6	62.1	59.3	60.3	56.7	0.3	9.6	16.1	14.6	26.5
North America	26	29	33	27	28	63.3	62.5	61.5	59.3	57.1	-0.4	-1.4	-1.3	-2.0	-3.3
EUROPE AND FORMER SOVIET UNION	**150**	**166**	**166**	**134**	**111**	**89.0**	**86.5**	**85.3**	**83.4**	**80.9**	**7.0**	**26.3**	**31.3**	**0.9**	**-15.5**
Former Soviet Union	82	89	84	63	42	115.3	112.1	109.6	107.3	102.9	8.5	18.7	18.3	-4.2	-19.6
Europe	68	77	82	72	69	75.3	73.8	73.6	71.6	70.4	-1.5	7.7	13.0	5.1	4.2

Note: World and regional figures include other regions and countries not reported separately. Country-level details are available online. Total production is aggregated across irrigated and rainfed systems at the national level and aligned with years as reported in FAOSTAT. Per capita food consumption is based on food availability at the national level. Net trade includes negative and positive numbers indicating that a region is a net importer or exporter, respectively, and balances to zero at the global level. Cereals include barley, millet, rice, sorghum, wheat, and aggregated other cereals. Meats include beef, pork, poultry, and sheep and goats. Fruits and vegetables include banana, plantain, aggregated temperate fruits, aggregated tropical fruits, and aggregated vegetables. Oilseeds include groundnuts, rapeseed, soybean, sunflower, and aggregated other oilseeds. Pulses include beans, chickpeas, cowpeas, lentils, pigeonpeas, and aggregated other pulses. Roots and tubers include cassava, potato, sweet potato, yams, and aggregated other roots and tubers. Values reported for 2010 are calibrated model results. Projections for 2030 and 2050 assume changes in population and income as reflected in the IPCC's Shared Socioeconomic Pathway 2. Climate change impacts are simulated using the IPCC's Representative Concentration Pathway 8.5 and the HadGEM general circulation model. Further documentation is available at www.ifpri.org/program/impact-model.

Source: IFPRI, IMPACT Model version 3.3, October 2016.

Notes

CHAPTER 1

1 FAO (Food and Agriculture Organization of the United Nations) et al., *The State of Food Security and Nutrition in the World, 2018: Building Climate Resilience for Food Security and Nutrition* (Rome: FAO, 2018).

2 FAO, *Asia and the Pacific Regional Overview of Food Security and Nutrition 2018—Accelerating Progress Towards the SDGs* (Bangkok: 2018); National Statistical Office and UNICEF, *Thailand 14 Provinces Multiple Indicator Cluster Survey 2015–2016, Final Report* (Bangkok: 2017).

3 FAO et al., *State of Food Security and Nutrition in the World, 2018.*

4 Ibid.

5 Ibid.

6 Poverty lines are calculated using 2011 PPP dollars. A. Castañeda et al., "A New Profile of the Global Poor," *World Development* 101 (2018): 250–267.

7 Development Initiatives, *2018 Global Nutrition Report: Shining a Light to Spur Action on Nutrition* (Bristol, UK: 2018).

8 UN (United Nations), *The Sustainable Development Goals Report 2018* (New York: 2018).

9 World Bank, *Global Economic Prospects, June 2018: The Turning of the Tide?* (Washington, DC: 2018).

10 Ibid.

11 F. Wouterse and A. S. Taffesse, eds., *Boosting Growth to End Hunger by 2025: The Role of Social Protection: ReSAKSS Annual Trends and Outlook Report 2017–2018* (Washington, DC: IFPRI, 2018).

12 USDA-FAS (United States Department of Agriculture–Foreign Agricultural Service), "World Agricultural Production," December 2018; OECD-FAO (Organisation for Economic Co-operation and Development-Food and Agricultural Organization of the United Nations), *Agricultural Outlook 2018–2027* (2018).

13 USDA-FAS, "World Agricultural Production."

14 OECD-FAO, "Agricultural Outlook 2018–2027."

15 FAO, Global Parliamentary Summit against Hunger and Malnutrition, held in Madrid, October 29–30, 2018, www.fao.org/about/meetings/global-parliamentary-summit/en/.

16 Global Panel on Agriculture and Food Systems for Nutrition, *Preventing Nutrient Loss and Waste across the Food System: Policy Actions for High-Quality Diets*, Policy Brief No. 12 (London: 2018).

17 T. Makombe, W. Tefera, and S. Benin, "Tracking Key CAADP Indicators and Implementation Processes," in *Boosting Growth to End Hunger by 2025*, ed. Wouterse and Taffesse.

18 A. Bossuyt, "Moving Towards Nutrition-Sensitive Agriculture Strategies and Programming: Ethiopia Country Case" in *Agriculture for Improved Nutrition: Seizing the Momentum*, ed. S. Fan, S. Yosef, and R. Pandya-Lorch (Wallingford, UK: CABI and IFPRI, 2019).

19 A. U. Ahmed and J. Ghostlaw, "Diversifying Rice-Centric Agriculture and Diets: The Bangladesh Experience," in *Agriculture for Improved Nutrition*, ed. Fan et al.

20 "Pakistan Launches Strategy to Address Nutrition Issue," *Daily Times*, May 17, 2018.

21 D. Wong and A. C. Koty, "The US-China Trade War: A Timeline," China Briefing, December 14, 2018.

22 K. Amado, "Trump's NAFTA Changes: The USMCA Would Create U.S. Jobs and Raise Auto Prices," *Balance*, December 4, 2018.

23 IFPRI (International Food Policy Research Institute), "China's Road to Food Security in a Time of Rising Trade Protectionism," press release, June 26, 2018.

24 World Bank, *Global Economic Prospects*, June 2018.

25 B. Girard, "What Will the US-China Trade War Mean for Africa?," *Diplomat*, October 25, 2018.

26 K. Bradsher, "Modi, in Davos, Praises Globalization without Noting India's Trade Barriers," *New York Times*, January 23, 2018.

27 D. K. Eberhart, "Trump's Trade Antics Are Encouraging Protectionism around the World," *Washington Examiner*, June 18, 2018.

28 Y. Yong, "RCEP Set for Early Conclusion," *Star* online, October 12, 2018.

29 M. Kennedy, "In Face of Protectionism, EU and Japan Sign Huge Open-Trade Deal," NPR, July 17, 2018.

30 African Union, "CFTA—Continental Free Trade Area," accessed January 9, 2019.

31 S. Edwards, "Donors Put Up $1B for Global Financing Facility, with Notable Absences," Devex, November 7, 2018.

32 B. Allen-Ebrahimian, "Treasury Takes Aim at Global Food Security Program," *Foreign Policy*, November 10, 2017.

33 UNCTAD (United Nations Conference on Trade and Development), *World Investment Report 2018: Investment and New Industrial Policies* (Geneva: United Nations, 2018).

34 "Africa Innovation Summit Kicks Off in Kigali," *East Africa Business Week Digital*, June 6, 2018.

35 "Investors to Explore US$82M Worth of Business Prospects at First Nutrition Investor Forum in Africa," *East Africa Business Week Digital*, October 18, 2018.

36 A. Saldinger et al., "UNGA Takeaways," Devex, October 1, 2018.

37 M. Allen et al., "Summary for Policymakers," in *Global Warming of 1.5°C: An IPCC Special Report on the Impacts of Global Warming of 1.5°C above Pre-industrial Levels and Related Global Greenhouse Gas Emission Pathways*, ed. V. Masson-Delmotte et al. (Geneva: Intergovernmental Panel on Climate Change, 2018).

38 CRED (Centre for Research on the Epidemiology of Disasters) and UNISDR (United Nations International Strategy for Disaster Reduction), *Economic Losses, Poverty, and Disasters: 1998–2017* (Brussels and Geneva: 2018).

39 FSIN (Food Security Information Network), *Global Report on Food Crises, 2018* (Rome: 2018).

40 Climate Action Tracker, "USA Country Summary," accessed January 9, 2019.

41 Climate Action Tracker, "Temperatures," accessed January 9, 2019.

42 FSIN, *Global Report on Food Crises 2018.*

43 S. Nebehay, "Yemeni Children Die as Warring Sides Block Aid Deliveries: UNICEF," Reuters, November 2, 2018; D. Walsh, "Yemen Cease-Fire Takes Effect: Why Now and What's Next?," *New York Times*, December 18, 2018.

44 M. Kohut and I. Herrera, "As Venezuela Collapses, Children Are Dying of Hunger," *New York Times*, December 17, 2017.

45 USB (Simón Bolívar University), UCV (Central University of Venezuela), and UCAB (Andrés Bello Catholic University), Survey on Living Conditions (2017); V. Sequera, "Venezuelans Report Big Weight Losses in 2017 as Hunger Hits," Reuters, February 21, 2018.

46 M. Specia and K. Bracken, "In South Sudan, a Never-Ending Hunger Season Puts Millions in Danger," *New York Times*, May 30, 2018.

47 World Food Programme, Counting the Beans: The True Cost of a Plate of Food Around the World, 2018, accessed January 25, 2019, https://cdn.wfp.org/2018/plate-of-food/.

48 UN, *World Economic Situation and Prospects, 2018* (New York: 2018).

49 Rabobank, "USDA WASDE–Wait and Xi," December 12, 2018.

50 World Bank, Global Economic Prospects, accessed January 10, 2019.

51 Scale Up Conference, University of Michigan College of Agriculture, https://ag.purdue.edu/scaleup/Pages/default.aspx.

TIMELINE

JANUARY: AU (African Union), "African Union Launches Africa Agriculture Transformation Scorecard (AATS)—A Revolutionary New Tool to Drive Agricultural Productivity and Development," January 29, 2018.

MARCH: S. Gustafson, "Conflict, Weather Shocks Driving Food Crises in 2017," Food Security Portal, March 26, 2018; L. Signe, "Africa's Big New Free Trade Agreement," *Washington Post*, March 29, 2018; AU, "CFTA—Continental Free Trade Area," accessed January 4, 2019.

JUNE: UN Department of Economic and Social Affairs, *The Sustainable Development Goals Report, 2018* (New York: UN, 2018); G7 Presidency (Canada), "The Whistler Principles to Accelerate Innovation for Development Impact," June 2, 2018.

JULY: G20 Information Centre, "Declaration: G20 Meeting of Agriculture Ministers," July 28, 2018; E. Callaway, "CRISPR Plants Now Subject to Tough GM Laws in European Union," *Nature*, July 25, 2018.

SEPTEMBER: FAO (Food and Agriculture Organization of the United Nations), "Global Hunger Continues to Rise, New UN Report Says," September 11, 2018; FAO et al., *The State of Food Security and Nutrition in the World 2018* (Rome: FAO, 2018); C. Shepherd and B. Blanchard, "China's Xi Offers Another $60 Billion to Africa, but Says No to 'Vanity Projects,'" Reuters, September 3, 2018; India, Ministry of Women and Child Development, "National Nutrition Month (Poshan Maah) Witnesses Overwhelming People's Participation," October 5, 2018.

OCTOBER: World Food Prize, "2018 Haddad and Nabarro," accessed January 4, 2019, https://www.worldfoodprize.org/index.cfm?nodeID=94236&audienceID=1; IPCC (Intergovernmental Panel on Climate Change), Global Warming of 1.5 °C, accessed January 4, 2019, https://www.ipcc.ch/sr15/; D. A. Weiss, "Global Food Security Act," Reuters, October 23, 2018; USAID (United States Agency for International Development), "Statement by Administrator Mark Green on the Reauthorization of the Global Food Security Act," October 12, 2018.

NOVEMBER: FAO and IFPRI, "Accelerating the End of Hunger and Malnutrition," http://www.ifpri-faobangkokconference.org/; Global Nutrition Report 2018, https://globalnutritionreport.org/.

DECEMBER: M. Igoe, "3 Takeaways from COP24," Devex, December 17, 2018; IPC (Integrated Food Security Phase Classification), "Yemen's Food Security Situation Remains Dire despite Substantial Humanitarian Assistance," December 7, 2018; M. Ryan, "'Immediate Responses' Are Needed to Save Lives in Yemen, U.N. and Aid Groups Say," *Washington Post*, December 10, 2018; D. Walsh, "Yemen Cease-Fire Takes Effect: Why Now and What's Next?," *New York Times*, December 18, 2018.

CHAPTER 2

1 IFAD (International Fund for Agricultural Development), *Rural Poverty Report 2011* (Rome: 2010).

2 S. Fan, A. Gulati, and S. Thorat, "Investment, Subsidies, and Pro-poor Growth in Rural India," *Agricultural Economics* 29 (2008): 163170; S. Fan, L. Zhang, and X. Zhang, "Reforms, Investment, and Poverty in Rural China," *Economic Development and Cultural Change* 52, no. 2 (2004): 395–421.

3 G. Thapa and G. Shively, "A Dose-Response Model of Road Development and Child Nutrition in Nepal," *Research in Transportation Economics* 70 (2018): 112–124.

4 UN (United Nations), *World Urbanization Prospects: The 2014 Revision* (New York: 2014).

5 OECD (Organisation for Economic Co-operation and Development), *Reinventing Rural Policy*, Policy Brief (Paris: 2006).

6 A. Odusola, "Agriculture, Rural Poverty and Income Inequality in Sub-Saharan Africa," in *Income Inequality Trends in Sub-Saharan Africa: Divergence, Determinants and Consequences*, ed. A. Odusola et al. (New York: UNDP Regional Bureau for Africa, 2017).

7 I. Boto et al., "Major Drivers for Rural Transformation in Africa: Resources on Rural Transformation in Africa," Brussels Rural Development Briefings 24, September 14, 2011.

8 B. Cohen, "Urbanization in Developing Countries: Current Trends, Future Projections, and Key Challenges for Sustainability," *Technology in Society* 28, no. 1-2 (2006): 63–80.

9 S. G. Rabinowicz and V. Chinapah, "Good Practices in Pursuit of Sustainable Rural Transformation," *Journal of Education and Research* 4, no. 2 (2014): 7–23.

10 S. Fan, J. Brzeska, and T. Olofinbiyi, "The Business Imperative: Helping Small Family Farmers to Move Up or Move Out," in *2014–2015 Global Food Policy Report* (Washington, DC: IFPRI, 2015).

11 D. Gollin, R. Jedwab, and D. Volrath, "Urbanization without Industrialization," *Journal of Economic Growth* 21, no. 1 (2015): 35–70.

12 P. A. Dorosh, E. Schmidt, and J. Thurlow, "Agricultural Growth, Poverty, and the Rural Nonfarm Economy: A Spatial Economy-Wide Analysis," in *Agriculture and the Rural Economy in Pakistan: Issues, Outlooks, and Policy Priorities*, ed. D. J. Spielman et al. (Philadelphia: University of Pennsylvania Press on behalf of IFPRI).

13 X. Diao et al., *Cities and Rural Transformation: A Spatial Analysis of Rural Youth Livelihoods in Ghana*, IFPRI Discussion Paper 1599 (Washington, DC: IFPRI, 2017).

14 A. M. Zu et al., *Rural Off-Farm Incomes in Myanmar's Dry Zone*, Food Security Policy Project Research Highlights (Washington, DC: IFPRI, 2017).

15 J. Janssens et al., *Mothers' Non-farm Entrepreneurship and Child Secondary Education in Rural Ghana*, IFPRI Discussion Paper 1705 (Washington, DC: IFPRI, 2018).

16 UNDP (United Nations Development Programme), *Commercializing Communities: Transitions in Water Management in Rural Senegal* (New York: 2013).

17 S. Khandker, Z. Bakht, and G. Koolwal, "The Poverty Impact of Rural Roads: Evidence from Bangladesh," *Economic Development and Cultural Change* 57, no. 4 (2009): 685–722.

18 UNDP, *Integrated Sustainable Rural Development: Renewable Energy Electrification and Rural Productivity Zones*, Discussion Paper (New York: UNDP, 2014).

19 E. Nakasone and M. Torero, "A Text Message Away: ICTs as a Tool to Improve Food Security," *Agricultural Economics* 47, no. S1 (2016): 49–59.

20 B. A. Abuya, J. Ciera, and E. Kimani-Murage, "Effect of Mother's Education on Child's Nutritional Status in the Slums of Nairobi," *BMC Pediatrics* 12 (2012): 80.

21 G. Psacharopoulos, *Benefits and Costs of the Education Targets for the Post-2015 Development Agenda: Post-2015 Consensus* (Tewksbury, MA, US: Copenhagen Consensus Center, 2014).

22 P. Nagler and W. Naudé, "Non-farm Entrepreneurship in Rural Sub-Saharan Africa: New Empirical Evidence," *Food Policy* 67 (February 2017): 175–191.

23 O. D. Anyamele, "Urban and Rural Differences across Countries in Child Mortality in Sub-Saharan Africa," *Journal of Health Care for the Poor and Underserved* 20, no. 4 (2009): 90–98.

24 R. Strasser, S. M. Kam, and S. M. Regalado, "Rural Health Care Access and Policy in Developing Countries," *Annual Review of Public Health* 37 (2016): 395–412.

25 R. Uauy et al., "Global Efforts to Address Severe Acute Malnutrition," *Journal of Pediatric Gastroenterology and Nutrition* 55, no. 5 (2012): 476–481; S. Guerrero and E. Rogers, "Is Community-Based Treatment of Severe Acute Malnutrition (SAM) at Scale Capable of Meeting Global Needs?," vol. 1, *Access for All* (London: Coverage Monitoring Network, 2013).

26 P. Dasgupta et al., "2014: Rural Areas," in *Climate Change 2014: Impacts, Adaptation, and Vulnerability,* pt. A of *Global and Sectoral Aspects*, ed. C. B. Field et al., Contribution of Working Group II to the Fifth Assessment Report of the Intergovernmental Panel on Climate Change (New York: Cambridge University Press, 2014).

27 IFAD, "Community-Based Integrated Natural Resources Management Project (CBINReMP): Supervision Report," February 2016.

28 Rabinowicz and Chinapah, "Good Practices in Pursuit of Sustainable Transformation"; IFAD, "Community-Based Integrated Natural Resources Management Project."

29 UNDP, *An Integrated Framework to Support Local Governance and Local Development* (New York: 2016).

30 J. Speer and W. Vásquez, *Opportunities and Challenges for Community Involvement in Public Service Provision in Rural Guatemala*, IFPRI Discussion Paper 1250 (Washington, DC: IFPRI, 2013).

31 S. Drimie and S. Yosef, "Reducing Risk, Strengthening Resilience: Social Protection and Nutrition," in *Nourishing Millions: Stories of Change in Nutrition*, ed. S. Gillespie et al. (Washington, DC: IFPRI, 2016).

32 UNDP, *Integrated Framework to Support Local Governance.*

33 FAO (Food and Agriculture Organization of the United Nations), *The State of Food and Agriculture: Leveraging Food Systems for Inclusive Rural Transformation* (Rome: 2017).

34 F. N. Bachewe et al., *Non-farm Income and Labor Markets in Rural Ethiopia*, ESSP II Working Paper 90 (Washington, DC, and Addis Ababa: IFPRI and Ethiopian Development Research Institute, 2016).

CHAPTER 3

1 Authors' calculations based on World Bank indicators SI.POV.RUHC; SI.POV.URHC; SP.RUR.TOTL; and SP.URB.TOTL, together with $1.90 national headcount rates calculated by authors using PovcalNet country-income distributions and household survey mean. The international poverty line of $1.90 per person per day (in 2011 purchasing power parity terms), defined by the World Bank, is based on 15 national poverty lines from some the poorest countries in the world. Initially set at $1 per person per day in 1990, the international poverty line was updated in 2008 to $1.25 and in 2015 to $1.90. The update accounts for changes in costs of living across the globe, reflecting changes in costs of basic food, clothing, and shelter needs globally.

2 Povcal.net, http://iresearch.worldbank.org/PovcalNet/povOnDemand.aspx, accessed December 2018.

3 FAO (Food and Agriculture Organization of the United Nations) et al., *The State of Food Security and Nutrition in the World 2018: Building Climate Resilience for Food Security and Nutrition* (Rome: FAO, 2018).

4 Ibid.

5 Ibid.

6 UNICEF, WHO, World Bank, Joint Child Malnutrition Estimates, 2018, https://www.who.int/nutgrowthdb/estimates/en/.

7 WHO, Global Health Observatory Data, accessed December 14, 2018, https://www.who.int/gho/child-malnutrition/en/.

8 ICF, DHS Program STATcompiler, http://www.statcompiler.com, accessed December 18, 2018.

9 Ibid.

10 L. C. Smith, H. Alderman, and D. Aduayom, Food Insecurity in Sub-Saharan Africa: New Estimates from Household Expenditure Surveys, Research Report 146 (Washington, DC: IFPRI, 2006); A. U. Ahmed et al., *The World's Most Deprived: Characteristics and Causes of Extreme Poverty and Hunger* (Washington, DC: IFPRI, 2007); H. Kismul et al., "Determinants of Childhood Stunting in the Democratic Republic of Congo: Further Analysis of Demographic and Health Survey, 2013-14," *BMC Public Health* 18, no. 1 (2017): 74.

11 H. Kharas, J. W. McArthur, and K. Rasmussen, *How Many People Will the World Leave Behind? Assessing Current Trajectories on the Sustainable Development Goals*, Global Economy and Development Working Paper 123 (Washington, DC: Brookings Institution, 2018).

12 J.-P. Robin, "Kristalina Georgieva: Global Warming Can Add 100 Million Poor People by 2030," World Bank, September 14, 2017.

13 World Bank, "Crisis Hitting Poor Hard in Developing World, World Bank Says," February 12, 2009.

14 J. Kwame Sundaram, "We Can Overcome Poverty and Hunger by 2030," Inter Press Service, September 29, 2015.

15 U. Lele, "Measuring Poverty and Hunger Can Raise More Questions Than Answers," Brookings Institution, June 15, 2015.

16 "For Up to 800 Million Rural Poor, a Strong World Bank Commitment to Agriculture," World Bank, November 12, 2014.

17 H. Kharas et al., *Ending Rural Hunger: Mapping Needs and Actions for Food Security and Nutrition* (Washington, DC: Brookings Institution, 2015). See also O. Ecker and C. Breisinger, *The Food Security System: A New Conceptual Framework* (Washington, DC: IFPRI, 2012).

18 World Bank, *World Development Report 2008: Agriculture for Development* (Washington, DC: 2007).

19 J. Lampietti, "Planet of the Apps: Making Small Farms Competitive," *World Bank Voices: Perspectives on Development,* October 15, 2018; M. W. Rosegrant et al., *Food Security in a World of Natural Resource Scarcity: The Role of Agricultural Technologies* (Washington, DC: IFPRI, 2014).

20 AgResults, accessed December 14, 2018, https://agresults.org/.

21 Kharas et al., *Ending Rural Hunger.*

22 Germany, Federal Ministry for Economic Cooperation and Development, "Tracking the L'Aquila Food Security Initiative Pledge and Related Funding," July 2015.

23 GAFSP (Global Agriculture and Food Security Program), Ending Poverty and Hunger, accessed December 14, 2018, http://web.archive.org/.

web/20180921085801/https://www.gafspfund.org/; US Department of the Treasury, "FY 2019 Budget in Brief," accessed December 14, 2018.

24 AgIncentives, "Nominal Rate of Protection," accessed December 14, 2018, http://www.ag-incentives.org/indicator/nominal-rate-protection.

25 K. A. Elliott, "The WTO, Agriculture, and Development: A Lost Cause?," International Centre for Trade and Sustainable Development, February 15, 2018.

26 WTO (World Trade Organization), "Kazakhstan to Host WTO's Next Ministerial Conference," July 26, 2018; WTO, "Goods and Services–What Is Being Traded?," in *World Trade Statistical Review 2018* (Geneva: 2018): 40-65.

27 African Union, "Africa Agriculture Transformation Scorecard: The 2017 Progress Report to the Assembly," 2018.

28 World Bank, Worldwide Governance Indicators Data Sources, accessed December 14, 2018, http://info.worldbank.org/governance/wgi/#doc-sources.

29 FAO, "Global Push to Stamp Out Hunger Hinges on Better Data," September 25, 2018.

30 "Draft Dubai Declaration: Supporting the Implementation of the Cape Town Global Action Plan for Sustainable Development Data," accessed December 14, 2018.

31 FAO, Save Food: Global Initiative on Food Loss and Waste Reduction, accessed December 14, 2018, http://www.fao.org/save-food/en/.

32 S. Jaffee et al., *The Safe Food Imperative: Accelerating Progress in Low and Middle-Income Countries*, Agriculture and Food Series (Washington, DC: World Bank, 2019).

33 H. Kharas, J. W. McArthur, and J. von Braun, "An Evidence-Based Approach to Ending Rural Hunger," *Economics E-Journal*, September 21, 2017.

34 Kharas et al., *Ending Rural Hunger.*

35 H. Yang, "Virtual Water Trade: An Assessment of Water Use Efficiency in the International Food Trade," *Hydrology and Earth System Sciences Discussions* 10, no. 3 (2006): 443-454; J. von Braun, A. Gulati, and H. Kharas, "Key Policy Actions for Sustainable Land and Water Use to Serve People," *Economics E-Journal*, November 2017.

36 W. Mwangi, "FABLE Consortium Launches Programme on Integrated Land-Use and Food Systems," International Institute for Sustainable Development, January 11, 2018.

37 von Braun et al. "Key Policy Actions."

38 WTO, "WTO Members Secure 'Historic' Nairobi Package for Africa and the World," December 19, 2015.

39 UN Sustainable Development Knowledge Platform, Voluntary National Reviews Database, accessed December 14, 2018, https://sustainabledevelopment.un.org/vnrs/.

40 S. Custer et al., *Listening to Leaders 2018: Is Development Cooperation Tuned-In or Tone-Deaf?* (Williamsburg, VA, US: AidData at the College of William & Mary, 2018).

41 GAFSP, Private Sector Window, accessed December 14, 2018, www.gafspfund.org/private-sector-window.

42 AlphaBeta, *Valuing the SDG Prize in Food and Agriculture: Unlocking Business Opportunities to Accelerate Sustainable and Inclusive Growth* (London: Business and Sustainable Development Commission, 2016).

43 S. Guarnaschelli et al., *Financing Sustainable Land Use: Unlocking Business Opportunities in Sustainable Land Use with Blended Finance* (London: Kois Invest, 2018).

44 AgResults, https://agresults.org/, accessed January 2, 2019.

CHAPTER 4

1 B. Davis, S. Di Giuseppe, and A. Zezza, "Are African Households (Not) Leaving Agriculture? Patterns of Households' Income Sources in Rural Sub-Saharan Africa," *Food Policy* 67 (2017): 153-174.

2 O. Badiane and T. Makombe, eds., *Beyond a Middle-Income Africa: Transforming African Economies for Sustained Growth with Rising Employment and Incomes*, ReSAKSS Annual Trends and Outlook Report 2014 (Washington, DC: IFPRI, 2015); D. Rodrik, "Premature Deindustrialization," *Journal of Economic Growth* 21, no. 1 (2016): 1-33; X. Diao, M. S. McMillan, and D. Rodrik, *The Recent Growth Boom in Developing Economies: A Structural-Change Perspective*, NBER Working Paper No. 23132 (Cambridge, MA: National Bureau of Economic Research, 2016).

3 Rodrik, "Premature Deindustrialization."

4 ILO (International Labour Organization), ILOSTAT database, accessed November 2018.

5 D. Gollin, R. Jedwab, and D. Vollrath, "Urbanization with and without Industrialization," *Journal of Economic Growth* 21, no. 1 (2016): 35-70.

6 M. McMillan, D. Rodrik, and C. Sepulveda, *Structural Change, Fundamentals, and Growth: A Framework and Country Studies* (Washington, DC: IFPRI, 2017).

7 J. Thurlow, "Youth Employment Prospects in Africa," in *African Youth and the Persistence of Marginalization: Employment, Politics, and Prospects for Change*, eds. D. Resnick and J. Thurlow (New York: Routledge, 2015).

8 Rodrik, "Premature Deindustrialization"; McMillan et al., *Structural Change, Fundamentals, and Growth.*

9 J. Thurlow, P. Dorosh, and B. Davies, "Demographic Change, Agriculture and Rural Poverty," in *Sustainable Food and Agriculture: An Integrated Approach*, ed. C. Campanhola and S. Pandey (London: Elsevier and Food and Agriculture Organization of the United Nations, 2019).

10 O. Badiane and J. M. Ulimwengu, "Business Pathways to the Future of Smallholder Farming in the Context of Transforming Value Chains," in *Africa Agriculture Status Report 2017: The Business of Smallholder Agriculture in Sub-Saharan Africa* (Nairobi: Alliance for a Green Revolution in Africa, 2017).

11 D. Byerlee et al., *Growing Africa - Unlocking the Potential of Agribusiness: Main Report* (Washington, DC: World Bank, 2013).

12 A. Mukherjee and X. Zhang, "Rural Industrialization in China and India: Role of Policies and Institutions," *World Development* 35: 1621-1634.

13 P. Hazell, S. Haggblade, and T. Reardon, "Structural Transformation of the Rural Non-farm Economy," in *Transforming the Rural Non-farm Economy*, ed. S. Haggblade, P. Hazell, and T. Reardon (Baltimore: Johns Hopkins University Press, 2007); FAO (Food and Agriculture Organization of the United Nations), "A Territorial Approach to Inclusive Rural Transformation," in *The State of Food and Agriculture, 2017: Leveraging Food Systems for Inclusive Rural Transformation* (Rome: 2017); A. Berdegué et al., "Cities, Territories, and Inclusive Growth: Unraveling Urban-Rural Linkages in Chile, Colombia, and Mexico," *World Development* 73 (2015): 56-71.

14 P. Dorosh and J. Thurlow, "Can Cities or Towns Drive African Development? Economywide Analysis for Ethiopia and Uganda," *World Development* 63 (2014): 113-123.

15 P. Dorosh, E. Schmidt, and A. Shiferaw, "Economic Growth without Structural Transformation: The Case of Ethiopia," *Journal of African Development* 14, no. 2 (2014): 7-40.

16 World Bank, World Development Indicators database, accessed November 2018, https://databank.worldbank.org/data/source/world-development-indicators .

17 Ibid.

18 P. Dorosh and J. Thurlow, "Agricultural Growth, Urbanization, and Poverty Reduction," in *Agricultural Development: New Perspectives in a Changing World*, ed. S. Fan and K. Otsuka (forthcoming).

19 Hazell et al., "Structural Transformation of the Rural Non-farm Economy."

20 T. Reardon, *Growing Food for Growing Cities: Transforming Food Systems in an Urbanizing World* (Chicago: Chicago Council on Global Affairs, 2016).

CHAPTER 5

1 For a systematic review of interventions increasing women's control of resources, see J. Yoong, L. Rabinovich, and S. Diepeveen, "The Impact of Economic Resource Transfers to Women versus Men: A Systematic Review," Technical Report (London: EPPI-Centre, Social Science Research Unit, Institute of Education, University of London, 2012). For studies on the relationship between women's empowerment and agricultural productivity, see G. Seymour, "Women's Empowerment in Agriculture: Implications for Technical Efficiency in Rural Bangladesh," *Agricultural Economics* 48, no. 4 (2017): 513–522; on the relationship between women's empowerment and household food security, see E. Sraboni et al., "Women's Empowerment in Agriculture: What Role for Food Security in Bangladesh?" *World Development* 61 (2014): 11–52; on the relationship between women's empowerment and maternal and child nutrition, see H. J. Malapit et al., "Women's Empowerment Mitigates the Negative Effects of Low Production Diversity on Maternal and Child Nutrition in Nepal," *Journal of Development Studies* 51, no. 8 (2015): 1097–1123.

2 M. T. Ruel and H. Alderman, "Nutrition-Sensitive Interventions and Programmes: How Can They Help to Accelerate Progress in Improving Maternal and Child Nutrition?," *Lancet* 382, no. 9891 (2013): 536–551.

3 S. K. Head et al., *Women's Lives and Challenges: Equality and Empowerment since 2000* (Rockville, MD: ICF International, 2014).

4 C. Doss et al., "Gender, Rural Youth, and Structural Transformation," background paper for 2019 International Fund for Agricultural Development Rural Development Report (forthcoming).

5 V. Slavchevska, S. Kaaria, and S.-L. Taivalmaa, *Feminization of Agriculture in the Context of Rural Transformations: What Is the Evidence?* (Washington, DC: World Bank, 2016).

6 N. Kabeer, "Resources, Agency, Achievements: Reflections on the Measurement of Women's Empowerment," *Development and Change* 30, no. 3 (1999): 435–464.

7 Slavchevska et al., *Feminization of Agriculture*.

8 H. J. Malapit et al., "Measuring Progress toward Empowerment: Women's Empowerment in Agriculture Index," Feed the Future/IFPRI/OPHI Report (Washington DC: IFPRI, 2014).

9 FAO (Food and Agriculture Organization of the United Nations), *The State of Food and Agriculture, 2010–11: Women in Agriculture—Closing the Gender Gap for Development* (Rome: 2011); T. Kilic, P. Winters, and C. Carletto, "Gender and Agriculture in Sub-Saharan Africa: Introduction to the Special Issue," *Agricultural Economics* 46, no. 3 (2015): 281–284.

10 D. Johnston et al., *Agriculture, Gendered Time Use, and Nutritional Outcomes: A Systematic Review*, IFPRI Discussion Paper 1456 (Washington, DC: IFPRI, 2015); M. van den Bold, A. R. Quisumbing, and S. Gillespie, *Women's Empowerment and Nutrition: An Evidence Review*, IFPRI Discussion Paper 1294 (Washington, DC: IFPRI, 2013).

11 P. Jagger and G. Shipley, "Land Use Change, Fuel Use and Respiratory Health in Uganda," *Energy Policy* 67 (April 2014): 713–726.

12 World Bank, *Gender and Governance in Rural Services: Insights from India, Ghana, and Ethiopia* (Washington DC: 2010).

13 C. Leisher et al., "Does the Gender Composition of Forest and Fishery Management Groups Affect Resource Governance and Conservation Outcomes? A Systematic Map," *Environmental Evidence* 5, no. 6 (2016).

14 O. Westermann, J. Ashby, and J. Pretty, "Gender and Social Capital: The Importance of Gender Differences for the Maturity and Effectiveness of Natural Resource Management Groups," *World Development* 33, no. 11 (2005): 1783–1799.

15 N. L. Johnson et al., "How Do Agricultural Development Projects Empower Women? Linking Strategies with Expected Outcomes," *Journal of Gender, Agriculture and Food Security* 3, no. 2 (2018): 1–19.

16 K. Davis et al., "Impact of Farmer Field Schools on Agricultural Productivity and Poverty in East Africa," *World Development* 40, no. 2 (2012): 402–413. Friis-Hansen and Duveskog also found that the farmer field schools contributed to farmers' empowerment, but they do not distinguish effects on male and female farmers. See E. Friis-Hansen and D. Duveskog, "The Empowerment Route to Well-Being: An Analysis of Farmer Field Schools in East Africa," *World Development* 40, no. 2 (2012): 414–427.

17 S. Alkire et al., "The Women's Empowerment in Agriculture Index," *World Development* 52 (2013): 71–79.

18 S. Theis et al., "What Happens after Technology Adoption? Gendered Aspects of Small-Scale Irrigation Technologies in Ethiopia, Ghana, and Tanzania," *Agriculture and Human Values* 35, no. 3 (2018): 671–684.

19 A. R. Quisumbing, N. Kumar, and J. A. Behrman, "Do Shocks Affect Men's and Women's Assets Differently? Evidence from Bangladesh and Uganda," *Development Policy Review* 36, no. 1 (2018): 3–34.

20 C. Manfre et al., *Reducing the Gender Gap in Agricultural Extension and Advisory Services: How to Find the Best Fit for Men and Women Farmers*, MEAS Discussion Paper (Urbana, IL: Modernizing Extension and Advisory Services, 2013); FAO, *The Gender and Rural Advisory Services Assessment Tool (GRAST)* (Rome: 2016).

21 Y. Hahn et al., "Education, Marriage, and Fertility: Long-Term Evidence from a Female Stipend Program in Bangladesh," *Economic Development and Cultural Change* 66, no. 2 (2018): 383–415.

22 N. Kumar and A. R. Quisumbing, "Policy Reform toward Gender Equality in Ethiopia: Little by Little the Egg Begins to Walk," *World Development* 67 (2015): 406–423.

23 N. Kumar and A. R. Quisumbing, "Beyond 'Death Do Us Part': The Long-Term Implications of Divorce Perceptions on Women's Well-Being and Child Schooling in Rural Ethiopia," *World Development* 40, no. 12 (2012): 2478–2489.

24 See L. Beaman et al., *Women Politicians, Gender Bias, and Policy-making in Rural India*, UNICEF Background Paper (New York: UNICEF, 2006). For empirical evidence on this issue and a review of other studies in India, see K. Raabe, M. Sekher, and R. Birner, *The Effects of Political Reservations for Women on Local Governance and Rural Service Provision—Survey Evidence for Karnataka*, IFPRI Discussion Paper 878 (Washington DC: IFPRI, 2009).

25 N. Kumar et al., "Social Networks, Mobility, and Political Participation: The Potential for Women's Self-Help Groups to Improve Access and Use of Public Entitlement Schemes in India," *World Development* 114 (2019): 28–41.

CHAPTER 6

1 M. Ives, "Outrage over Fish Kill in Vietnam Simmers 6 Months Later," *New York Times*, October 30, 2016; M. Lopes, "Activists Feared Brazil's Bolsonaro Would Accelerate Amazon Deforestation. Now They Think It's Already Happening," *Washington Post*, December 7, 2018; M. Allen et al., "Summary for Policymakers," in *Global Warming of 1.5°C: An IPCC Special Report on the Impacts of Global Warming of 1.5°C above Pre-industrial Levels and Related Global Greenhouse Gas Emission Pathways*, ed. V. Masson-Delmotte et al. (Geneva: Intergovernmental Panel on Climate Change, 2018).

2 FAO (Food and Agriculture Organization of the United Nations) et al., *The State of Food Security and Nutrition in the World 2018: Building Climate Resilience for Food Security and Nutrition* (Rome: FAO, 2018).

3 CGIAR Research Program on Water, Land and Ecosystems (WLE), *Ecosystem Services and Resilience Framework* (Colombo: International Water Management Institute, WLE, 2014).

4 Our definition of sustainable rural areas is built on a definition of sustainable agriculture from the FAO. Sustainable rural systems are those that (1) are managed toward maintaining ecosystem functions for present and future needs, (2) support rural livelihoods, and (3) provide economic benefits for everyone. See FAO, *Building a Common Vision for Sustainable Food and Agriculture: Principles and Approaches* (Rome: 2014); N. Desai and C. Ringler, "Green Economy: Sustainable and Growing, but Also Food Secure?" in *2012 Global Food Policy Report* (Washington, DC: IFPRI, 2013).

5 B. W. Griscom et al., "Natural Climate Solutions," *PNAS* 114, no. 44 (2017): 11645–11650.

6 X.-P. Song et al., "Global Land Change from 1982 to 2016," *Nature* 560 (2018): 639–643.

7 FAO and ITPS (Intergovernmental Technical Panel on Soils), *Status of the World's Soil Resources (SWSR)—Main Report* (Rome: FAO, 2015).

8 H. Xie and C. Ringler, "Agricultural Nutrient Loadings to the Freshwater Environment: The Role of Climate Change and Socioeconomic Change," *Environmental Research Letters* 12, no. 10 (2017).

9 M. Wang et al., "Rural Industries and Water Pollution in China," *Journal of Environmental Management* 86 (2008): 648–659; China Water Risk, "2016 State of Environment Report Review," June 14, 2017.

10 S. Siebert et al., "Groundwater Use for Irrigation—A Global Inventory," *Hydrology and Earth System Sciences* 14, no. 10 (2010): 1863–1880.

11 S. Kaza et al., *What a Waste 2.0: A Global Snapshot of Solid Waste Management to 2050: Trends in Solid Waste Management* (Washington, DC: World Bank, 2018).

12 S. E. Bauer, K. Tsigaridis, and R. Miller, "Significant Atmospheric Aerosol Pollution Caused by World Food Cultivation," *Geophysical Research Letters* 43, no. 10 (2016): 5394–5400.

13 S. Bonjour et al., "Solid Fuel Use for Household Cooking: Country and Regional Estimates for 1980–2010," *Environmental Health Perspectives* 121, no. 7 (2013): 784–790; WHO (World Health Organization), *7 Million Premature Deaths Annually Linked to Air Pollution* (Geneva: 2014).

14 M. Subramanian, "Delhi's Deadly Air: India's Capital Scrambles to Tackle Its Epic Pollution Problems," *Nature* 534 (2016): 166–169.

15 R. S. Ferguson and S. T. Lovell, "Permaculture for Agroecology: Design, Movement, Practice, and Worldview: A Review," *Agronomy for Sustainable Development* 34, no. 2 (2014): 251.

16 P. Kristjanson et al., "Addressing Gender in Agricultural Research for Development in the Face of a Changing Climate: Where Are We and Where Should We Be Going?," *International Journal of Agricultural Sustainability* 15, no. 5 (2017): 482–500.

17 R. McLain et al., "Toward a Tenure-Responsive Approach to Forest Landscape Restoration: A Proposed Tenure Diagnostic for Assessing Restoration Opportunities," *Land Use Policy* (forthcoming).

18 C. Briceño-Garmendia, A. Estache, and N. Shafik, *Infrastructure Services in Developing Countries: Access, Quality, Costs and Policy Reform*, Policy Research Working Paper 3468 (Washington, DC: World Bank, 2004).

19 W. S. Shin et al., "Forest Experience and Psychological Health Benefits: The State of the Art and Future Prospect in Korea," *Environmental Health and Preventive Medicine* 15, no. 1 (2010): 38–47.

20 R. Q. Grafton, J. Horne, and S. A. Wheeler, "On the Marketisation of Water: Evidence from the Murray-Darling Basin, Australia," *Water Resources Management* 30, no. 3 (2016): 913–926.

21 M. W. Rosegrant et al., *Food Security in a World of Natural Resource Scarcity: The Role of Agricultural Technologies* (Washington, DC: IFPRI, 2014).

22 N. Lefore et al., "Sustainable and Equitable Growth in Farmer-Led Irrigation in Sub-Saharan Africa: What Will It Take?," *Water Alternatives* (forthcoming).

23 D. Higgins et al., "Investigating the Impacts of Increased Rural Land Tenure Security: A Systematic Review of the Evidence," *Journal of Rural Studies* 61 (2018): 34–62.

24 D. A. Ali, K. Deininger, and M. Goldstein, "Environmental and Gender Impacts of Land Tenure Regularization in Africa: Pilot Evidence from Rwanda," *Journal of Development Economics* 110 (2014): 262–275; K. Deininger et al., "Rural Land Certification in Ethiopia: Process, Initial Impact, and Implications," *World Development* 36, no. 10 (2008): 1786–1812.

25 N. Kumar and A. R. Quisumbing, "Policy Reform toward Gender Equality in Ethiopia: Little by Little the Egg Begins to Walk," *World Development* 67 (2015): 406–423; K. Deininger, D. A. Ali, and T. Yamano, "Legal Knowledge and Economic Development: The Case of Land Rights in Uganda," *Land Economics* 84, no. 4 (2008): 593–619.

26 R. Meinzen-Dick, Q. Bernier, and E. Haglund, *The Six "Ins" of Climate-Smart Agriculture: Inclusive Institutions for Information, Innovation, Investment, and Insurance*, CAPRi Working Paper No. 114 (Washington, DC: IFPRI, 2013).

27 L. Persha and C. Meshack, *A Triple Win? The Impact of Tanzania's Joint Forest Management Programme on Livelihoods, Governance and Forests*, 3ie Impact Evaluation Report 34 (New Delhi: International Initiative for Impact Evaluation, 2016); T. Blomley et al., "Seeing the Wood for the Trees: An Assessment of the Impact of Participatory Forest Management on Forest Condition in Tanzania," *Oryx* 42, no. 3 (2008): 380–391; N. Gross-Camp, "Tanzania's Community Forests: Their Impact on Human Well-Being and Persistence in Spite of the Lack of Benefit," *Ecology and Society* 22, no. 1 (2017): 37.

CHAPTER 7

1 World Bank, *State of Electricity Access Report (SEAR), 2017* (Washington, DC: 2018).

2 IEA (International Energy Agency) et al., *Tracking SDG7: The Energy Progress Report* (Washington, DC: World Bank, 2018).

3 The levelized cost of energy allows simple comparisons of costs of different methods of electricity generation. It is the average total cost per unit of electricity generated to build and operate a power-generating asset divided by the total energy output over the lifetime of that asset. The levelized cost can also be understood as the average minimum price at which electricity must be sold to break even. Limitations of levelized cost calculations are addressed later in the main text.

4 B. Wehrmann, "Solar Power in Germany—Output, Business and Perspectives," Fact Sheet, Clean Energy Wire, September 22, 2018.

5 Committee on Climate Change, "Reducing UK Emissions: 2018 Progress Report to Parliament," June 2018.

6 J. Wright et al., *Formal Comments on the Integrated Resource Plan (IRP) Update Assumptions, Base Case and Observations* (Pretoria: Council for Scientific and Industrial Research, 2017); T. J. Reber et al., "Preliminary Findings of the South Africa Power System Capacity Expansion and Operational Modelling Study: Preprint," Office of Scientific and Technical Information, US Department of Energy, January 1, 2018; C. Arndt et al., *Developments in Variable Renewable Energy and Implications for Developing Countries*,SA-TIED Working Paper 6 (UN-WIDER, 2018).

7 Returning to Figure 1 and considering the period between 2014 and 2017, the cost of commercial PV systems fell by about 33 percent, driven by reductions in costs for all elements of the system.

8 B. Baulch, T. D. Do, and T. H. Le, "Constraints to the Uptake of Solar Home Systems in Ho Chi Minh City and Some Proposals for Improvement," *Renewable Energy* 118 (2018): 245–256.

9 World Bank, *State of Electricity Access Report (SEAR) 2017.*

10 D. Headey, J. Hoddinott, and K. Hirvonen, *Animal Sourced Foods and Child Stunting*, IFPRI Discussion Paper 1695 (Washington, DC: IFPRI, 2017).

11 H. Alderman, J. Hoddinott, and B. Kinsey, "Long Term Consequences of Early Childhood Malnutrition," *Oxford Economic Papers* 58, no. 3 (2006): 450–474; J. Hoddinott et al., "Effect of a Nutrition Intervention during Early Childhood on Economic Productivity in Guatemalan Adults," *Lancet* 371, no. 9610 (2008): 411–416.

12 J. A. Burney, R. L. Naylor, and S. L. Postel, "The Case for Distributed Irrigation as a Development Priority in Sub-Saharan Africa," *PNAS* 110, no. 31 (2013): 12513–12517; H. Xie et al., "Estimating the Potential for Expanding Smallholder Irrigation in Sub-Saharan Africa," *Agricultural Water Management* 131 (2014): 183–193; L. You et al., "What Is the Irrigation Potential for Africa? A Combined Biophysical and Socioeconomic Approach," *Food Policy* 36, no. 6 (2011): 770–782.

13 S. Passarelli et al., "Evaluating the Pathways from Small-Scale Irrigation to Dietary Diversity: Evidence from Ethiopia and Tanzania," *Food Security* 10, no. 4 (2018): 981–997.

14 S. Theis et al., "What Happens after Technology Adoption? Gendered Aspects of Small-Scale Irrigation Technologies in Ethiopia, Ghana, and Tanzania," *Agriculture and Human Values* 35, no. 3 (2018): 671–684.

15 Theis et al., "What Happens after Technology Adoption?"; Burney et al., "The Case for Distributed Irrigation"; IRENA, "Solar Pumping for Irrigation: Improving Livelihoods and Sustainability," IRENA Policy Brief, 2016.

16 L. Nigussie et al., *Gender and Water Technologies: Water Lifting for Irrigation and Multiple Purposes in Ethiopia*, Africa RISING Report (Addis Ababa: International Livestock Research Institute, 2017).

17 IEA et al., *Tracking SDG7.*

18 Ibid.

CHAPTER 8

1 AGRA (Alliance for a Green Revolution in Africa), *Africa Agriculture Status Report 2018: Catalyzing Government Capacity to Drive Agricultural Transformation* (Nairobi: 2018); USAID (US Agency for International Development), *U.S. Government Global Food Security Strategy FY2017-2021* (Washington, DC: 2016).

2 R. Rotberg, "Good Governance Means Performance and Results," *Governance* 27, no. 3 (2014): 511–518.

3 K. Deininger, *Land Policies for Growth and Poverty Reduction* (Washington, DC, and Oxford, UK: World Bank and Oxford University Press, 2003); K. Deininger, S. Jin, and H. K. Nagarajan, "Land Reforms, Poverty Reduction, and Economic Growth: Evidence from India," *Journal of Development Studies* 45, no. 4 (2009): 496–521; S. Holden, K. Deininger, and H. Ghebru, "Impacts of Low-Cost Land Certification on Investment and Productivity," *American Journal of Agricultural Economics* 91, no. 2 (2009): 359–373; S. Lawry et al., "The Impact of Land Property Rights Interventions on Investment and Agricultural Productivity in Developing Countries: A Systematic Review," *Campbell Systematic Reviews* 10, no. 1 (2014); K. Deininger and D. A. Ali, "Do Overlapping Land Rights Reduce Agricultural Investment? Evidence from Uganda (2008-03)," *American Journal of Agricultural Economics* 90, no. 4 (2008): 869–882.

4 G. Feder et al., *Land Policies and Farm Productivity in Thailand* (Baltimore: Johns Hopkins University Press, 1988); T. Besley, "Property-Rights and Investment Incentives—Theory and Evidence from Ghana," *Journal of Political Economy* 103, no, 5 (1995): 903–937; M. Carter and P. Olinto, "Getting Institutions 'Right' for Whom? Credit Constraints and the Impact of Property Rights on the Quantity and Composition of Investment," *American Journal of Agricultural Economics* 85, no. 1 (2003): 173–186.

5 K. Kosec and O. Shemyakina, "Land Reform and Child Health in the Kyrgyz Republic" (unpublished 2019).

6 World Bank, *The Ethiopian Urban Migration Study 2008: The Characteristics, Motives and Outcomes of Migrants to Addis Ababa*, Technical Report No. 55731-ET (Washington, DC: 2010); G. Zeleke et al., *Seasonal Migration and Rural Livelihoods in Ethiopia: An Empirical Study*, Working Paper Series on Rural-Urban-Linkage Theme of the Global Mountain Program, Working Paper 3 (Lima, Peru: Global Mountain Program, 2008); P. Dorosh et al., "Road Connectivity, Population, and Crop Production in Sub-Saharan Africa," *Agricultural Economics* 43, no. 1 (2012): 89–103.

7 K. Deininger, S. Jin, and H. K. Nagarajan, *Efficiency and Equity Impacts of Rural Land Rental Restrictions: Evidence from India* (Washington, DC: World Bank, 2007).

8 M. McMillan, D. Rodrik, and C. Sepúlveda, eds., *Structural Change, Fundamentals, and Growth: A Framework and Case Studies* (Washington, DC: IFPRI, 2017); C. Newman et al., *Made in Africa: Learning to Compete in Industry* (Washington, DC: Brookings Institution, 2016).

9 For more information, see http://eba.worldbank.org/about-us.

10 K. Kosec, "The Child Health Implications of Privatizing Africa's Urban Water Supply," *Journal of Health Economics* 35 (2014): 1–19.

11 S. Galiani, P. Gertler, and E. Schargrodsky, "Water for Life: The Impact of the Privatization of Water Services on Child Mortality," *Journal of Political Economy* 113 (2005): 83–120.

12 A. Banerjee et al., "Private Outsourcing and Competition: Subsidized Food Distribution in Indonesia," *Journal of Political Economy* (forthcoming).

13 S. Haggblade and A. Diarra, *Trends in West African Pesticide Markets*, Feed the Future Innovation Lab for Food Security Policy, Policy Research Brief No. 49, (East Lansing, MI: Michigan State University, 2017).

14 D. Resnick, "The Politics of Crackdowns on Africa's Informal Workers," *Comparative Politics* (forthcoming, 2019).

15 M. Andrews, L. Pritchett, and M. Woolcock, *Building State Capability: Evidence, Analysis, Action* (Oxford: Oxford University Press, 2017); World Bank, *World Development Report: Governance and the Law* (Washington, DC: 2017).

16 M. Lipsky, *Street-Level Bureaucracy: Dilemmas of the Individual in Public Services* (New York: Russell Sage Foundation, 1980).

17 C. Prendergast, "Intrinsic Motivation and Incentives," *American Economic Review* 98, no. 2 (2008): 201–205.

18 M. Andrews, L. Pritchett, and M. Woolcock, "Escaping Capability Traps through Problem Driven Iterative Adaptation (PDIA)," *World Development* 51 (2013): 234–244.

19 M. Yé et al., "Health Worker Preferences for Performance-Based Payment Schemes in a Rural Health District in Burkina Faso," *Global Health Action* 9, no. 1 (2016): 1–12.

20 Z. Hasnain, N. Manning, and J. Pierskalla, "The Promise of Performance Pay? Reasons for Caution in Policy Prescriptions in the Core Civil Services," *World Bank Research Observer* 29, no. 2 (2014): 235–264.

21 D. Gilligan et al., *Educator Incentives and Educational Triage in Rural Primary Schools*, NBER Working Paper No. 24911 (Cambridge, MA: National Bureau of Economic Research, 2018).

22 E. Dal Bó, F. Finan, and M. Rossi, "Strengthening State Capabilities: The Role of Financial Incentives in the Call to Public Services," *Quarterly Journal of Economics* 128, no. 3 (2013): 1169–1218.

23 M. Mauzinzi and J. Rubyutsa, "When Tradition Feeds on Modern Accountability Mechanisms in Public Policy Implementation: The Case of 'Imihigo' in Rwanda," *Public Performance and Management Review* 36, no. 3 (2018): 1–25.

24 K. Ggome and R. Newfarmer, *From Devastation to Services-First Transformation*, WIDER Working Paper 2017/84 (Helsinki: UNU-WIDER, 2017).

25 R. Shostak et al., *When Might the Introduction of a Delivery Unit Be the Right Intervention?* Governance and Public Sector Management Practice Notes (Washington, DC: World Bank, 2014).

26 C. Sabel and L. Jordan, *Doing, Learning, Being: Some Lessons Learned from Malaysia's National Transformation Program* (Washington, DC: World Bank, 2015).

27 W. Oates, "Searching for Leviathan: An Empirical Study," *American Economic Review* 75, no. 4 (1985): 748–757.

28 D. Brinkerhoff and O. Azfar, "Decentralization and Community Empowerment," in *Making Decentralization Work: Democracy, Development, and Security*, ed. E. Connerly, K. Eaton, and P. Smoke (Boulder, CO: Lynne Rienner, 2010).

29 A. Channa and J. P. Faguet, "Decentralization of Health and Education in Developing Countries: A Quality-Adjusted Review of the Empirical Literature," *World Bank Research Observer* 31, no. 2 (2016): 199–241.

30 D. Resnick, *The Devolution Revolution: Implications for Agricultural Service Delivery in Ghana*, IFPRI Discussion Paper 1714 (Washington, DC: IFPRI, 2018).

31 T. Mogues and T. Olofinbiyi, "Budgetary Influence under Information Asymmetries: Evidence from Nigeria's Subnational Agricultural Investments" (unpublished 2018).

32 E. Mwambulukutu, I. Minde, and D. Nyange, "Tax Reforms Benefit Farmers and Traders: A Case of Tea Production and Marketing in Tanzania," Michigan State University, Feed the Future Innovation Lab for Food Security Policy, September 26, 2018.

33 J. Fearon, "Electoral Accountability and the Control of Politicians: Selecting Good Types versus Sanctioning Poor Performance," in *Democracy, Accountability, and Representation*, ed. A. Przeworski, S. C. Stokes, and B. Manin (Cambridge: Cambridge University Press, 1999); T. Besley, *Principled Agents? The Political Economy of Good Government* (Oxford: Oxford University Press, 2006); J. W. Hatfield and G. Padró i Miquel, "A Political Economy Theory of Partial Decentralization," *Journal of the European Economic Association* 10, no. 3 (2012): 605–633.

34 D. Stasavage, "Democracy and Education Spending in Africa," *American Political Science Review* 49, no. 2 (2005): 343–358; R. H. Bates and S. A. Block, "Revisiting African Agriculture: Institutional Change and Productivity Growth," *Journal of Politics* 75, no. 2 (2013): 372–384.

35 T. Jayne et al., "Taking Stock of Africa's Second-Generation Agricultural Input Subsidy Programs," *Food Policy* 75 (February 2018): 1–14.

36 L. Blaydes and M. Kayser, "Counting Calories: Democracy and Distribution in the Developing World," *International Studies Quarterly* 55 (2011): 887–908; K. Kosec et al., "Political Competition and Rural Welfare: Evidence from Pakistan," *Oxford Economic Papers* 70, no 4. (2018): 1036–1061.

37 Kosec et al., "Political Competition and Rural Welfare."

38 G. Grossman and J. Lewis, "Administrative Unit Proliferation," *American Political Science Review* 108, no. 1 (2014): 196–217; C. Boone and M. Wahman, "Rural Bias in African Electoral Systems: Legacies of Unequal Representation in African Democracies," *Electoral Studies* 40 (2015): 335–346.

39 D. Resnick, "Democracy, Decentralization, and District Proliferation: The Case of Ghana," *Political Geography* 59 (2017): 47–60.

40 J. Gottlieb and K. Kosec, "The Countervailing Effects of Competition on Public Goods Provision: When Bargaining Inefficiencies Lead to Bad Outcomes," *American Political Science Review* 113, no. 1 (2019): 88—107.

41 A. Banerjee et al., "Tangible Information and Citizen Empowerment: Identification Cards and Food Subsidy Programs in Indonesia," *Journal of Political Economy* 126, no. 2 (2018): 451–491.

42 J. C. S. Serrato and P. Wingender, "Estimating the Incidence of Government Spending" (unpublished 2011); K. Kosec, "Relying on the Private Sector: The Income Distribution and Public Investments in the Poor," *Journal of Development Economics* 107 (2014): 320–342.

43 World Bank, World Development Indicators, accessed 2017, https://data.worldbank.org/data-catalog/world-development-indicators.

44 J. C. Aker and I. M. Mbiti, "Mobile Phones and Economic Development in Africa," *Journal of Economic Perspectives* 24, no. 3 (2010): 207–232; J. A. Ferejohn and J. H. Kuklinski, *Information and Democratic Processes* (Champaign, IL: University of Illinois Press, 1990); M. Fey, "Stability and Coordination in Duverger's Law: A Formal Model of Pre-election Polls and Strategic Voting," *American Political Science Review* 91, no. 1 (1997): 135–147; P. Keefer and S. Khemani, "Democracy, Public Expenditures, and the Poor: Understanding Political Incentives for Providing Public Services," *World Bank Research Observer* 20, no. 1 (2005): 1–27; T. Fujiwara and L. Wantchekon, "Can Informed Public Deliberation Overcome Clientelism? Experimental Evidence from Benin," *American Economic Journal: Applied Economics* 5, no. 4 (2013): 241–255.

45 T. Besley and R. Burgess, "The Political Economy of Government Responsiveness: Theory and Evidence from India," *Quarterly Journal of Economics* 117, no. 4 (2002): 1415–1451; D. Strömberg, "Radio's Impact on Public Spending," *Quarterly Journal of Economics* 119 (2004): 189–221.

46 R. Jia, M. Kudamatsu, and D. Seim, "Political Selection in China: The Complementary Roles of Connections and Performance," *Journal of the European Economic Association* 13, no. 4 (2015): 631–668.

47 N. Karachiwalla et al., "Promotion Incentives for Improved Service Delivery" (unpublished 2018); E. Duflo, R. Hanna, and S. P. Ryan, "Incentives Work: Getting Teachers to Come to School," *American Economic Review* 102, no. 4 (2012): 1241–1278.

48 J. C. Aker, P. Collier, and P. C. Vicente, "Is Information Power? Using Mobile Phones and Free Newspapers during an Election in Mozambique," *Review of Economics and Statistics* 99, no. 2 (2017): 185–200; G. Grossman, M. Humphreys, and G. Sacramone-Lutz, "'I wld like u WMP to extend electricity 2 our village': On Information Technology and Interest Articulation," *American Political Science Review* 108, no. 3 (2014): 688–705.

49 K. Kosec and L. Wantchekon, "Can Information Improve Governance and Rural Service Delivery?," *World Development* (forthcoming, 2019).

50 C. Cruz, P. Keefer, and J. Labonne, *Incumbent Advantage, Voter Information and Vote Buying*, IDB Working Paper Series 711 (Washington, DC: Inter-American Development Bank, 2016).

CHAPTER 9

1 This regional typology is based on the share of the regional population living in rural grid cells, also called the rural population. According to the definition, the predominantly urban regions have less than 20 percent rural population, intermediate regions between 20 and 50 percent, and predominantly rural regions more than 50 percent. Taken together, the intermediate regions and rural regions account for 55 percent of the EU's population and 90 percent of its territory.

2 EC (European Commission), "The Cork Declaration—A Living Countryside," November 11, 1996.

3 DG AGRI (Directorate-General for Agriculture and Rural Development), *Technical Handbook on the Monitoring and Evaluation Framework of the Common Agricultural Policy 2014–2020* (Brussels: European Commission, DG AGRI, 2017).

4 See DG AGRI Evaluation Reports: Rural Development, available at https://ec.europa.eu/agriculture/evaluation/rural-development-reports_en.

5 A. Andersson, S. Höjgård, and E. Rabinowicz, *Evaluation of Results and Adaptation of EU Rural Development Programmes*, Working Paper 2017:1 (Lund: University of Lund, Agri-food Economics Centre, 2017).

6 European Court of Auditors, *Can the Commission and Member States Show That the EU Budget Allocated to the Rural Development Policy Is Well Spent?*, Special Report No. 12 (Luxembourg: 2013); European Court of Auditors, *Rural Development Programming: Less Complexity and More Focus on Results Needed*, Special Report No. 16 (Luxembourg: 2017).

7 EC, *A Sustainable Bioeconomy for Europe: Strengthening the Connection between Economy, Society and the Environment: Updated Bioeconomy Strategy* (Brussels: 2018).

8 European Court of Auditors, *Can the Commission and Member States Show That the EU Budget Allocated to the Rural Development Policy Is Well Spent?*

9 Eurostat, Agri-Environmental Indicator—Commitments, Statistics Explained, June 2017.

10 EC, *Report from the Commission to the European Parliament and the Council on the Implementation of the Common Monitoring and Evaluation Framework and First Results on the Performance of the Common Agricultural Policy*, COM (2018) 790 (Brussels: 2018). It is likely that the discrepancy in these figures arises because the Eurostat figure refers to 2013, which is the previous programming period, and the Commission figures refer to plans for the 2014–2020 period.

11 Royal Society for the Protection of Birds/Birdlife International, *Seeds of Success: How Agri-Environment Can Yield Results for Nature and Farming* (Sandy, UK: RSPB, 2011).

12 European Court of Auditors, *Implementation of the LEADER Approach for Rural Development*, Special Report No. 5 (Luxembourg: 2010).

13 EC, *EU Action for Smart Villages* (Brussels: 2017).

14 R. Jongeneel, *Research for AGRI Committee—The CAP Support beyond 2020: Assessing the Future Structure of Direct Payments and the Rural Developments Interventions in the Light of the EU Agricultural and Environmental Challenges* (Brussels: European Parliament, Policy Department for Structural and Cohesion Policies, 2018).

REGIONAL DEVELOPMENTS

AFRICA

1 IMF (International Monetary Fund), *Regional Economic Outlook: Sub-Saharan Africa: Multispeed Growth* (Washington, DC: 2016); IMF, *World Economic Outlook: Challenges to Steady Growth* (Washington, DC: 2018).

2 Unless otherwise stated, figures in this section refer to Africa at the continent level and are based on data from T. Makombe, W. Tefera, and S. Benin, "Tracking Key CAADP Indicators and Implementation Processes," in *Boosting Growth to End Hunger by 2025: The Role of Social Protection: 2017–2018 ReSAKSS Annual Trends and Outlook Report*, ed. F. Wouterse and A. S. Taffesse (Washington, DC: IFPRI, 2018).

3 O. Badiane and M. McMillan, "Economic Transformation in Africa: Patterns, Drivers, and Implications for Future Growth Strategies," in *Beyond a Middle Income Africa: Transforming African Economics for Sustained Growth with Rising Employment and Incomes: 2014 ReSAKSS Annual Trends and Outlook Report,* ed. O. Badiane and T. Makombe (Washington, DC: IFPRI, 2015).

4 IFPRI, microsimulation modeling results for Kenya and Côte d'Ivoire, available from authors.

5 Based on data from World Bank, World Development Indicators database, accessed November 29, 2018, http://databank.worldbank.org/data.

6 World Bank, World Development Indicators.

7 "Ethiopia-Eritrea Border Reopens after 20 Years," BBC, September 11, 2018.

8 "South Sudan's President, Rebel Leader Sign Peace Deal," Reuters, September 12, 2018.

9 UNCTAD (United Nations Conference on Trade and Development), *World Investment Report 2018: Investment and New Industrial Policies* (Geneva: 2018).

10 IMF, *World Economic Outlook: Challenges to Steady Growth*.

11 B. Girard, "What Will the US-China Trade War Mean for Africa?," *Diplomat*, October 25, 2018.

12 IMF, *Regional Economic Outlook: Sub-Saharan Africa: Cash Flows and the Future of Work* (Washington, DC: 2018).

13 AU (African Union), "Agreement Establishing the African Continental Free Trade Area," 2018.

14 AU, CFTA-Continental Free Trade Area, accessed November 1, 2018, https://au.int/en/ti/cfta/about.

15 Trade Law Centre (Tralac), "African Continental Free Trade Area (AfCFTA) Legal Texts and Policy Documents," 2019, accessed February 20, 2019.

16 "Cautious Nigeria Agrees to Sign African Continental Free Trade Agreement," Business Live, July 12, 2018.

17 United Nations Economic Commission for Africa, "Giant Trade Agreement for Development in Africa: The Continental Free Trade Area," Africa Trade Policy Center, Addis Ababa, 2018.

18 Ibid.

19 D. Tschirley, S. Haggblade, and T. Reardon, "Africa's Emerging Food System Transformation," White Paper for USAID, Michigan State University, Project of the Global Center for Food System Innovation and the Food Security Policy Innovation Lab, 2013.

20 V. Thériault et al., *A City-Retail Outlet Inventory of Processed Dairy and Grain Foods: Evidence from Mali*, Feed the Future Innovation Lab for Food Security Policy Research Paper 65 (East Lansing: Michigan State University, 2017); R. Andam et al., *Is Ghana Making Progress in Agro-processing? Evidence from an Inventory of Processed Food Products in Retail Shops in Accra*, Ghana Strategy Support Program Working Paper 41 (Washington, DC: IFPRI, 2015); J. Snyder et al., *Stages of Transformation in Food Processing and Marketing: Results of an Initial Inventory of Processed Food Products in Dar es Salaam, Arusha, and Mwanza*, Food Security Policy Innovation Lab Tanzania Policy Research Brief 3 (Dar es Salaam: 2015).

21 O. Badiane and J. M. Ulimwengu, "Business Pathways to the Future of Smallholder Farming in the Context of Transforming Value Chains," in *Africa Agriculture Status Report 2017: The Business of Smallholder Agriculture in Sub-Saharan Africa* (Nairobi: AGRA, 2017).

22 D. Tschirley et al., "Africa's Unfolding Diet Transformation: Implications for Agrifood System Employment," *Journal of Agribusiness in Developing and Emerging Economies* 5, no.1 (2015).

23 F. Hollinger and J. Staatz, *Agricultural Growth in West Africa: Market and Policy Drivers* (Rome: FAO, African Development Bank, and ECOWAS, 2015).

24 C. Cazzuffi, M. Pereira-López, and I. Soloaga, "Local Poverty Reduction in Chile and Mexico: The Role of Food Manufacturing Growth," *Food Policy* 68 (2017): 160–185; F. J. Proctor and J. A. Berdegué, *Food Systems at the Rural-Urban Interface*, Working Paper Series 194 (Santiago, Chile: Rimisp, 2016).

25 T. Sonobe and K. Otsuka, *Cluster-Based Industrial Development—A Comparative Study of Asia and Africa* (London: Palgrave Macmillan, 2011); O. Badiane and M. S. McMillan, "Economic Transformation in Africa: Patterns, Drivers, and Implications for Future Growth Strategies," in *Beyond a Middle Income Africa*, ed. Badiane and Makombe.

26 A. Adelaja et al., "A Policy Agenda for Achieving an Inclusive Transformation of Africa's Agri-food Systems," in *Africa Agriculture Status Report 2017*.

27 M. Torero, "Infrastructure Challenges and Potential for Africa South of the Sahara," in *Beyond a Middle Income Africa*, ed. Badiane and Makombe.

28 Hollinger and Staatz, *Agricultural Growth in West Africa*.

29 B. Handjiski, "Mobile Connectivity in Africa Has Already Arrived," Brookings Future Development Blog, Brookings Institution, March 18, 2015.

30 T. Makombe, W. Tefera, and S. Benin, "Tracking Key CAADP Indicators and Implementation Processes," in *Boosting Growth to End Hunger by 2025*, ed. Wouterse and Taffesse.

31 AfDB (African Development Bank Group), "Programme for Infrastructure Development in Africa (PIDA)"; African Development Bank Group, "The 2018 Programme for Infrastructure Development in Africa Week Opens with Calls for Good Governance to Accelerate Africa's Integration through Smart Infrastructure," November 27, 2018.

32 AfDB, "African Development Bank to Play Bigger Role in African Economies by Investing in Infrastructure Private Equity," November 16, 2018.

33 B. McKibben, "The Race to Solar-Power Africa," *New Yorker*, June 26, 2017.

34 L. You et al., "What Is the Irrigation Potential for Africa? A Combined Biophysical and Socioeconomic Approach," *Food Policy* 36, no. 6 (2011): 770–782.

35 IFPRI, "The Impact of Research on the Productive Safety Net Programme," ESSP Outcome Note 5, IFPRI, Addis Ababa, 2014.

36 M. Reimers and S. Klasen, "Revisiting the Role of Education for Agricultural Productivity," *American Journal of Agricultural Economics 95*, no. 1 (2011): 131–152; S. Allen et al., "Government Expenditures, Health Outcomes, and Marginal Productivity of Agricultural Inputs: The Case of Tanzania," *Journal of Agricultural Economics* 65, no. 3 (2014), 637–661.

37 IMF, *World Economic Outlook*.

MIDDLE EAST AND NORTH AFRICA

1 United States Energy Information Agency, "Short-Term Energy Outlook," October 10, 2018.

2 OECD/FAO (Organisation for Economic Co-operation and Development/ Food and Agriculture Organization of the United Nations), *OECD-FAO Agricultural Outlook 2018-2027* (Paris and Rome: 2018); FAO, Global Information and Early Warning System (GIEWS) Country Briefs for Yemen, Syria, Libya, and Iraq, June 2017.

3 "Syria War: Why Does the Battle for Idlib Matter?" BBC, September 17, 2018.

4 "Yemen at 'Crucial Moment,' Says Special Envoy, Stressing Need for Political Approach to Ending Conflict," United Nations Security Council, UNSC 8404th meeting, November 16, 2018.

5 "Food Markets in the Time of Conflict and Cholera: Rapid Market Assessment in Yemen," WFP (World Food Programme), April 2018.

6 IMF (International Monetary Fund), "MENAP Oil-Importing Countries: Safeguarding the Growth Recovery amid Rising Risks," in *Regional Economic Outlook: Middle East and Central Asia* (Washington, DC: 2018).

7 EIU (Economist Intelligence Unit), Country Report for Jordan, August 2018.

8 World Bank, *A New Economy for the Middle East and North Africa* (Washington, DC: 2018).

9 EIU Country Reports for Kuwait and Oman, August 2018.

10 ILO (International Labour Organization), *Global Employment Trends for Youth 2015: Scaling Up Investments in Decent Jobs for Youth* (Geneva: 2015).

11 EIU Country Report for Syria, August 2018.

12 World Bank, *A New Economy for the Middle East and North Africa*.

13 R. Dimova, S. Elder, and K. Stephan, *Labor Market Transitions of Young Women and Men in the Middle East and North Africa* (Geneva: ILO, 2016).

14 A. Nin Pratt et al., *Agriculture and Economic Transformation in the Middle East and North Africa: A Review of the Past with Lessons for the Future* (Washington, DC, and Rome: IFPRI and FAO, 2018).

15 World Bank, FAO, and International Fund for Agricultural Development, *Improving Food Security in Arab Countries* (Washington, DC: World Bank, 2009).

16 For example, IMF "Unlocking the Potential of People in MENA," Infographic, January 2018.

17 K. M. Dashash, "The Government Role in Creating Innovation Technological Clusters in the Developing Countries (The Case of Saudi Arabia)" (PhD Diss., University of Massachusetts-Amherst, 2017).

18 J. Ruan and X. Zhang, *Made in China: Crisis Begets Quality Upgrade*, IFPRI Discussion Paper 1025 (Washington, DC: IFPRI, 2009); F. Abdelaziz et al., *Clusters as Drivers of Local Industrial Development in Egypt: Which Are the Promising Sectors and Locations?* MENA RP Working Paper 10 (Washington, DC: IFPRI, 2018).

19 Abdelaziz et al., *Clusters as Drivers of Local Industrial Development*.

20 Ibid.

21 UN ESCWA (United Nations Economic and Social Commission for Western Asia), "Beirut Consensus on Technology for Sustainable Development in the Arab Region: Technology and Innovation for Decent Employment and Empowered Youth," Resolutions and Outcome Document Adopted at the Thirtieth Session of ESCWA, Beirut, June 2018.

22 UN ESCWA, "Regional Initiative for Promoting Small-Scale Renewable Energy Applications in Rural Areas of the Arab Region," Beirut, September 2018.

23 IMF, "Unlocking the Potential of People in MENA."

24 OECD, *Dialogue Maroc-OCDE sur les politiques de développement territorial: Enjeux et recommandations pour une action publique coordonnée* (Paris: 2018).

25 Agrytech, "About the Agrytech Program," accessed December 11, 2018, http://agrytech.org/about/.

CENTRAL ASIA

1 IMF (International Monetary Fund), *Regional Economic Outlook: Middle East and Central Asia* (Washington, DC: October 2018).

2 World Bank, "Commodity Markets Outlook, Oil Exporters: Policies and Challenges," April 2018; authors' calculations using data from various sources.

3 Central Bank of the Russian Federation, Database on Exchange Rates, accessed December 28, 2018, http://cbr.ru/currency_base/; and National Bank of Kazakhstan, Official Exchange Rates Database, accessed December 28, 2018.

4 Central Bank of the Russian Federation, "Statistical Data on Cross-Border Transfers of Individuals," accessed December 28, 2018, https://www.cbr.ru/Eng/statistics/?PrtId=svs.

5 IMF, "Regional Economic Outlook: Middle East and Central Asia," October 2018.

6 R. Satke, "Understanding Turkmenistan's Food Shortages," *Diplomat*, December 13, 2018.

7 ADB (Asian Development Bank), "Asian Economic Integration Report 2018: Toward Optimal Provision of Regional Public Goods in Asia and the Pacific," October 2018.

8 A. Yuldashev, "Tajikistan Ratified Agreement on Visa-Free Travel with Uzbekistan," *Asia-Plus*, March 19, 2018.

9 Uzbekistan, State Committee on Statistics, "Foreign Trade Turnover of the Republic of Uzbekistan," January-September 2018.

10 China, National Bureau of Statistics, "2017 China Statistical Yearbook."

11 Uzbekistan, State Committee on Statistics, accessed October 2018, https://stat.uz/uz/press-sluzhba/novosti-gks/4899-tashqi-savdo-8.

12 World Bank, "Turkmenistan Country Snapshot," October 2018.

13 L. Burunciuc et al.,"How Uzbekistan Is Transforming into an Open Economy," Future Development blog, Brookings Institution, December 20, 2018.

14 Tajikistan Statistical Agency, Ministry of Health and Social Protection of Population of the Republic of Tajikistan, and ICF, *Tajikistan Demographic and Health Survey 2017: Key Indicators* (Rockville, MD: 2018); R. Lavado, W. Seitz, and A. Thiebaud, *Childhood Stunting in Tajikistan: Quantifying the Association with WASH, Food Security, Health, and Care Practices* (Washington, DC: World Bank, 2017).

15 Turkmenistan, State Committee of Statistics and UNICEF, *2015-2016 Turkmenistan Multiple Indicator Cluster Survey, Final Report* (Ashgabat: 2016).

16 UNICEF/WHO/World Bank Group, *Levels and Trends in Child Malnutrition: Key Findings of the 2018 Edition of the Joint Child Malnutrition Estimates* (2018).

17 FAO, *The State of Food Security and Nutrition in Europe and Central Asia, 2017* (Rome: 2017).

18 Our calculations show significant and persistent productivity gaps between the agriculture and nonagriculture sectors. In Kazakhstan, agriculture employs 15.4 percent of the employed, but it accounts for only 4.4 percent of GDP. Similarly, agriculture employed 26.2, 27.3, and 64.1 percent of the working population and produced 12.3, 17.0, and 21.1 percent of GDP in 2017, in Kyrgyzstan, Uzbekistan, and Tajikistan, respectively.

19 K. A. Nevmatulina, "Role of Special Economic Zones in Development of the Republic of Kazakhstan," *Middle-East Journal of Scientific Research 15*, no. 11 (2013): 1528-1532.

20 ADB, "A Diagnostic Study of Kazakhstan's Special Economic Zones and Industrial Zones," April 2018.

21 ADB, "Strategic Framework for Special Economic Zones and Industrial Zones in Kazakhstan," May 2018.

22 Uzbekistan, Office of the President, Presidential Decree Nos. 5388, 3843, 3573, and 3657, Tashkent, Uzbekistan, 2018; Uzbekistan, Cabinet, Cabinet Ministers' Resolution No. 752, Tashkent, Uzbekistan, 2018.

23 Uzbekistan, Ministry of Foreign Affairs, "Effectiveness of the Planned Tasks Implementation for Economic Development," September 13, 2018.

24 World Bank, "Doing Business 2019: Training for Reform," October 2018.

25 K. Anderson et al., eds., *Kazakhstan: Accelerating Economic Diversification* (Manila: ADB, 2018); KazAgroFinance, "KAF Has Launched the Active Spring Field Campaign," accessed December 3, 2018.

26 ADB, "Country Operations Business Plan: Kazakhstan 2019-2021," September 2018; "Country Operations Business Plan: Kyrgyz Republic 2019-2012," October 2018; "Country Operations Business Plan: Uzbekistan 2018-2020," October 2017.

27 World Bank, "World Bank Group President Meets Uzbek President, Pledges Support for Uzbekistan's Reforms," press release, May 17, 2018.

28 Burunciuc et al.,"How Uzbekistan Is Transforming."

29 IMF, *Regional Economic Outlook: Middle East and Central Asia.*

30 F. de Soyres et al., *How Much Will the Belt and Road Initiative Reduce Trade Costs?* World Bank Policy Research Paper 8614 (Washington, DC: World Bank, 2018).

SOUTH ASIA

1 World Bank, World Bank Open Data, accessed December 10, 2018, https://data.worldbank.org/.

2 R. Chand and S. K. Srivastava, "Changes in the Rural Labour Market and Their Implications for Agriculture," *Economic and Political Weekly* 49, no. 10 (2014): 47-54; S. Pramanik, D. Uttam, and C. Bantilan, "Rural Non-farm Economy in Bangladesh: Nature, Extent, Trends and Determinants," paper presented at the 8th Conference of the Asian Society of Agricultural Economists, BRAC Centre for Development Management, Savar, Dhaka, Bangladesh, October 15-17, 2014.

3 World Bank, World Bank Open Data.

4 Afghanistan, Central Statistics Organization, *National Risk and Vulnerability Assessment 2011-2012*, Afghanistan Living Conditions Survey (Kabul: 2014); Bangladesh, Bureau of Statistics, Ministry of Planning; and UNESCO, *Literacy Assessment Survey 2008* (Dhaka: 2008); Bhutan, NSB (National Statistics Bureau) and World Bank, *Bhutan Living Standards Survey Report, 2017* (Thimphu: 2017); India, Ministry of Statistics and Programme Implementation, "Literacy and Education," *Women and Men in India—2017* (New Delhi: 2017); Maldives, Ministry of Finance and Treasury,

Maldives–Population and Housing Census, 2014 (Malé: National Bureau of Statistics, 2015); Pakistan, Ministry of Finance, "Education," in *Pakistan Economic Survey 2016–17* (Islamabad: 2017); Sri Lanka, Ministry of Policy Planning and Economic Affairs, Department of Census and Statistics, Census of Population and Housing, 2012 (Battaramulla, Sri Lanka: 2015).

5 World Bank, World Bank Open Data.

6 UNICEF, South Asia Headline Results, 2018–2021, accessed January 30, 2019, www.unicefrosa-progressreport.org/stopstunting.html.

7 Development Initiatives, *2018 Global Nutrition Report: Shining a Light to Spur Action on Nutrition* (Bristol, UK: 2018).

8 FAO (Food and Agriculture Organization of the United Nations), *Dynamic Development, Shifting Demographics, Changing Diets* (Bangkok: 2018).

9 Bangladesh, National Institute of Population Research and Training, Ministry of Health and Family Welfare, Mitra and Associates, and ICF International, *Bangladesh Demographic and Health Survey, 2014* (Dhaka, Bangladesh, and Rockville, MD, US: 2016); Bhutan, NSB, *Bhutan Multiple Indicator Survey Report, 2011* (Thimphu: 2011); European Commission, "Country Profile on Nutrition: Afghanistan," July 2017; India, Ministry of Health and Family Welfare, *National Family Health Survey-4*, India Fact Sheet (Mumbai: International Institute for Population Sciences, 2016); Maldives, Ministry of Health and Family and ICF Macro Calverton, *Maldives Demographic and Health Survey, 2009* (Malé, Maldives, and Calverton, MD, US: 2010); Nepal, Ministry of Health, New ERA, and ICF, *Nepal 2016 Demographic and Health Survey: Key Findings* (Kathmandu: Nepal, Ministry of Health, 2017); Pakistan, National Institute of Population Studies and ICF, *Pakistan Demographic and Health Survey, 2017–18* (Islamabad, Pakistan, and Rockville, MD, US: 2018); Sri Lanka, Department of Census and Statistics, "Nutrition of Children and Women," in *Sri Lanka Demographic and Health Survey, 2016* (Colombo: 2017).

10 World Bank, Migration and Remittances Data, accessed December 20, 2018, www.worldbank.org/en/topic/migrationremittancesdiasporaissues/brief/migration-remittances-data.

11 World Bank, World Bank Open Data.

12 B. Gadde et al., "Air Pollutant Emissions from Rice Straw Open Field Burning in India, Thailand and the Philippines," *Environmental Pollution* 157 (2009): 1554–1558.

13 India, Ministry of Finance, "Budget Speech 2018–19."

14 Pakistan, Ministry of Finance, "Budget Speech 2018–19."

15 Bangladesh, Ministry of Finance, "Budget Speech 2018–19."

16 Nepal, Ministry of Finance, "Budget Speech of Fiscal Year 2018/19."

17 Sri Lanka, Ministry of Finance, "Budget Speech 2018."

EAST AND SOUTHEAST ASIA

1 World Bank, *"Navigating Uncertainty" East Asia and Pacific Economic Update (October)* (Washington, DC: 2018).

2 OECD (Organisation for Economic Co-operation and Development), *Economic Outlook for Southeast Asia, China and India 2018: Fostering Growth through Digitalisation* (Paris: 2018).

3 FAO (Food and Agriculture Organization of the United Nations) et al., *The State of Food Security and Nutrition in the World 2018* (Rome: FAO, 2018).

4 D. Eckstein, V. Künzel, and L. Schäfer, *Global Climate Risk Index 2018: Who Suffers Most from Extreme Weather Events? Weather-Related Loss Events in 2016 and 1997 to 2016* (Bonn: Germanwatch, 2017).

5 FAO , Global Information and Early Warning System, "Country Briefs," accessed July 13, 2018.

6 Philippines, Climate Change Commission, *National Climate Change Action Plan, 2011–2028*; Myanmar, Ministry of Agriculture and Irrigation, *Myanmar: Climate-Smart Agriculture Strategy*; Cambodia, National Climate Change Committee, *Cambodia Climate Change Strategic Plan 2014–2023*; Indonesia, Ministry of National Development Planning/National Development Planning Agency, *National Action Plan for Climate Change Adaptation (RAN-API)*; Viet Nam, Ministry of Agriculture and Rural Development, *Action Plan on Climate Change Response of Agriculture and Rural Development Sector in the Period 2011–2015 and Vision to 2050*; China, National Development and Reform Commission, *National Climate Change Adaptation Strategy*.

7 M. W. Rosegrant et al., "Climate Change and Asian Agriculture," *Asian Journal of Agriculture and Development* 7, no. 1 (2010): 41–81; "More Research Needed for Climate Resilience," *Vietnamnet*, November 30, 2018; G. Vulturius and M. Boyland, "Can Insurance Help Southeast Asia's Farmers Cope with Climate Change?," Stockholm Environment Institute, October 18, 2016; Agroinsurance, "Malaysia: Crop Insurance Scheme in Final Stage, Ministry Hopes to Make It for 2018 Budget," March 29, 2017; S. Cheng, "Crop Insurance Takes Root," *Phnom Penh Post*, March 1, 2016; T. Ko Ko, "Crop Insurance to Protect Paddy Farmers Available Soon," *Myanmar Times*, April 20, 2018; S. Hallegatte et al., *Shock Waves: Managing the Impacts of Climate Change on Poverty* (Washington, DC: World Bank, 2016); J. Gonsalves et al., eds., *Towards Climate Resilience in Agriculture for Southeast Asia: An Overview for Decision-Makers* (Hanoi: International Center for Tropical Agriculture, CGIAR Research Program on Climate Change, Agriculture and Food Security, 2015); FAO, *Dynamic Development, Shifting Demographics, Changing Diets* (Bangkok: 2018); FAO, "ASEAN Working to Build Resilience through Disaster-Responsive Social Protection Programmes," August 31, 2018.

8 CAAS (Chinese Academy of Agricultural Sciences), *Agriculture Sector Development Report* (Beijing: 2018); "Indonesia's Rice Self-Sufficiency to Be Achieved by 2020," *Jakarta Post*, March 30, 2017; "Malaysia Targets 80% Rice Self-Sufficiency by 2022," *Sun Daily*, August 29, 2018; P. Ranada, "PH Still Gunning for Rice Self-Sufficiency despite Duterte Remarks," *Rappler*, June 14, 2018.

9 FAO, *Food Price Monitoring and Analysis #7* (Rome: 2018); "Philippines' Duterte Approves 'Unimpeded' Rice Importation," Xinhua, October 9, 2018; C. O. Avendaño, "With Duterte Directive, Anyone Can Import Rice," *Inquirer.net*, October 10, 2018; K. R. Ocampo, "NFA to Control Rice Imports," *Inquirer.net*, October 17, 2018; C. Elemia, "Senate Approves Bill Lifting Rice Import Limits," *Rappler*, November 14, 2018.

10 C. P. Timmer, "Sowing the Seeds for Better Food Policy in Asia," East Asia Forum, July 26, 2018; FAO et al., *State of Food Security*; C. Maitra, *A Review of Studies Examining the Link between Food Insecurity and Malnutrition*, technical paper (Rome: FAO, 2018).

11 ASEAN (Association of Southeast Asian Nations), UNICEF, and WHO, *Regional Report on Nutrition Security in ASEAN, Volume 2* (Bangkok: UNICEF, 2016).

12 K. Chen and Z. Wang, "Agriculture and Nutrition in China," in *Agriculture for Improved Nutrition: Seizing the Momentum*, ed. S. Fan, S. Yosef, and R. Pandya-Lorch (Wallingford, UK: CABI, 2019); ASEAN, *ASEAN Regional Guidelines on Food Security and Nutrition Policy* (Jakarta: 2017).

13 C. P. Bown, E. Jung, and Z. Lu, "Trump and China Formalize Tariffs on $260 Billion of Imports and Look Ahead to Next Phase," Peterson Institute for International Economics, September 20, 2018.

14 China and the United States reached an agreement on a 90-day truce after the G20 meeting, with serious negotiations to resolve the tariff dispute to start immediately. China agreed to purchase substantial amounts of agricultural products from the US, but no details (or orders) have been forthcoming. The immediate sense of relief in global markets ended quickly and has given way to considerable skepticism that any agreement is imminent.

15 J. Mason, H. Gu, and K. Plume, "Inside China's Strategy in the Soybean Trade War," Reuters, December 27, 2018; W. Shen, "Trade Row Prompts China to Use Low-Protein Diet for Pigs to Reduce Soybean Intake," *Global Times*, August 6, 2018; H. Liu, "China Will Reduce Millions of Soybean Imports in 2018 (in Chinese)," *Xinhua*, August 6, 2018; Chinese Ministry of Agriculture and Rural Affairs, "2018 Preferential Policy for Farmers," April 3, 2018; CAAS, *Agriculture Sector Development Report*.

16 USDA-FAS (United States Department of Agriculture Foreign Agricultural Service), "China Announces Tariff Reductions for Asia-Pacific Trade Agreement Members," June 28, 2018; S. Chin, "Trade War: Will ASEAN Be Next?," *ASEAN Post*, July 16, 2018; "Singapore, Malaysia Could Be Most Exposed to US-China Trade War: OCBC," *Business Times*, July 11, 2018; "What Opportunities Is the US-China Trade War Likely to Bring for Vietnam?," *Vietnamnet*, July 21, 2018.

17 P. Arunmas, "Despite Seed Concerns, CPTPP Study to Proceed," *Bangkok Post*, October 1, 2018; "The Livestock Sector is Heavily Influenced by CPTPP," *Customersnews*, March 26, 2018.

18 ASEAN, ASEAN Economic Community 2025 Consolidated Strategic Action Plan, August 14, 2018; ASEAN, "ASEAN Updates AEC 2025 Consolidated Strategic Action Plan (CSAP)," August 14, 2018.

19 J. Y. Kwon, "Cost-Benefit Analysis of Japan's Decision to Join CPTPP," *Asia Times*, October 21, 2018.; "CPTPP Likely to Be Approved in Mid-November," *Vietnamnet,* October 25, 2018; Y. Yong, "RCEP Set for Early Conclusion," *Star Online,* October 12, 2018.

20 Chinese State Council, *Strategic Plan for Rural Revitalization (2018–2022)*; "Over 4000 Billion Yuan Will Be Invested in Rural Infrastructure Projects," Economic Information Daily, October 23, 2018; "China Plans USD 700 bln Investment into Rural Infrastructure," *China Knowledge*, October 24, 2018; "China Eyes More Rural Infrastructure Investment with PPPs," *China Daily*, February 18, 2017; T. Wu, "China Telecom to Invest 1.5 Billion Yuan in Rural Areas," *China News*, September 28, 2018.

21 Myanmar, Ministry of Agriculture, Livestock, and Irrigation, "Myanmar Agriculture Development Strategy and Investment Plan (2018-19~2022-23)"; Cambodia, Ministry of Agriculture, Forestry and Fisheries, "Agricultural Sector Strategic Development Plan 2014-2018"; IFAD, "Review of Experience of the National Target Program for New Rural Development" (2016); N. T. Kien and B. Minh, "Nation Target Programme on New Rural Development: The Evolution of Concepts in Historical Perspective," *Vietnam Social Sciences* 6 (2015): 41-55.

22 C. S. Choe, "Key Factors to Successful Community Development: The Korean Experience," Institute of Developing Economies-Japan External Trade Organization, Chiba, Japan, 2005; Korea Saemaul Undong Center, accessed October 20, 2018, https://www.saemaul.or.kr/eng/; C. Denpaiboon and K. Amatasawatdee, "Similarity and Difference of One Village One Product (OVOP) for Rural Development Strategy in Japan and Thailand," *Japanese Studies Journal Special Issue: Regional Cooperation for Sustainable Future in Asia* (2012): 52-62; K. Natsuda et al., *One Village One Product–Rural Development Strategy in Asia: The Case of OTOP in Thailand*, Working Paper, Ritsumeikan Center for Asia Pacific Studies (Beppu, Japan: Ritsumeikan Asia Pacific University, 2011); S. Wancharoen, "Villages Go All Out to Charm Tourists," *Bangkok Post*, May 15, 2018.

23 "Thailand Pivots Digital Strategies and Seeks to Grow Tech Sector with an Impact on GDP," *Asia One*, September 21, 2017; Malaysia, Economic Planning Unit, Ministry of Economic Affairs, "Eleventh Malaysia Plan, 2016-2020"; Chinese State Council, "Internet Plus Agriculture Model to Promote Integrated Rural Development," June 27, 2018.

24 Devdiscourse, "ADB's Fresh Support to Cambodia in Climate Resilience and Agriculture Sector," July 4, 2018; F. Arkin, "Asian Countries Need Not Fear of Automation," Science and Development Network, May 2, 2018; "Interview: Belt and Road Initiative Provides Opportunities for FAO to Collaborate with China, Says FAO Chief," *Xinhua*, April 15, 2018.

25 L. Liu, "E-commerce Revitalizes China's Rural Areas," *People's Daily*, March 14, 2018; Y. Ono, "Alibaba Expands Cross-Border E-commerce in Southeast Asia," *Nikkei Asia Review*, April 20, 2018; T. Takahashi, "Alibaba a Game Changer for Southeast Asian Development," *Nikkei Asia Review*, June 7, 2018; Z. Soo and C. Chen, "JD.com Hastens E-commerce Race in Southeast Asia with US$550 Million Stake Sale to Google," *South China Morning Post*, June 19, 2018.

LATIN AMERICA AND THE CARIBBEAN

1 FAO (Food and Agriculture Organization of the United Nations) et al., *2018 Panorama de la seguridad alimentaria y nutricional en América Latina y el Caribe* (Santiago, Chile: 2018).

2 A. Bouët and D. Laborde Debucquet, *US Trade Wars with Emerging Countries in the 21st Century: Make America and Its Partners Lose Again*, IFPRI Discussion Paper 1669 (Washington, DC: IFPRI, 2017); D. Laborde and V. Piñeiro, *Argentina and a Global Trade System in Turmoil* (Buenos Aires: Centro de Estudios Economicos, forthcoming).

3 E. Díaz-Bonilla et al., "Potential Impact of Changes in U.S. Migration Policies on Poverty and Food Security in Central American Countries" (unpublished, IFPRI, 2018; forthcoming as a Technical Paper, International Centre for Trade and Sustainable Development, Geneva).

4 IMF (International Monetary Fund), República Bolivariana de Venezuela, accessed October 2018, https://www.imf.org/en/Countries/VEN.

5 M. Landaeta-Jiménez et al., *ENCOVI (Encuesta sobre Condiciones de Vida Venezuela) 2017* (Caracas: Universidad Católica Andrés Bello, 2018).

6 This section draws on E. Díaz-Bonilla and R. G. Echeverría, "Duality, Urbanization, and Modernization of Agri-food Systems in Latin America and the Caribbean," in *Agricultural Development: New Perspectives in a Changing World*, ed. K. Otsuka and S. Fan (forthcoming).

7 A. de Janvry and E. Sadoulet, "Rural Poverty in Latin America," *Food Policy* 25, no. 4 (2000): 389-409; A. Schejtman and J. Berdegué, "Desarrollo territorial rural," *Debates y temas rurales* 1 (Santiago, Chile: RIMISP, 2004).

8 T. Reardon and J. A. Berdegué, "The Rapid Rise of Supermarkets in Latin America: Challenges and Opportunities for Development," *Development Policy Review* 20, no. 4 (2002): 371-388.

9 E. Díaz-Bonilla and M. Torero, "Regional Developments: Latin America," in *2016 Global Food Policy Report* (Washington, DC: IFPRI, 2016).

10 C. Bolling and M. Gehlhar, "Global Food Manufacturing Reorients to Meet New Demands," in *New Directions in Global Food Markets*, ed. A. Regmi and M. Gehlhar (Washington, DC: US Department of Agriculture/Economic Research Service, 2005).

11 T. Reardon and P. Timmer, "The Economics of the Food System Revolution," *Annual Review of Resource Economics* 4, no. 1 (2012): 225-264.

12 Reardon and Berdegué, "Rapid Rise of Supermarkets in Latin America."

13 Reardon and Timmer, "The Economics of the Food System Revolution."

FOOD POLICY INDICATORS: TRACKING CHANGE

AGRICULTURAL TOTAL FACTOR PRODUCTIVITY (TFP)

1 The data and the methodology used to build the database are available at: https://www.ers.usda.gov/data-products/international-agricultural-productivity/.

Made in the USA
Las Vegas, NV
04 November 2026

57431640R00100